Roman Imperial Artillery

Archaeopress Roman Archaeology 114

Roman Imperial Artillery

Outranging the enemies of the empire

Alan Wilkins

Archaeopress Archaeology

Archaeopress Publishing Ltd
Summertown Pavilion
18-24 Middle Way
Summertown
Oxford OX2 7LG
www.archaeopress.com

ISBN 978-1-80327-783-7
ISBN 978-1-80327-784-4 (e-Pdf)

© Alan Wilkins and Archaeopress 2024

First edition Shire 2008
Second edition Solway Print Dumfries 2017
This third edition Archaeopress 2024

The *ballista* stone-thrower, reconstructed from Vitruvius' text and Heron's advice and diagram (Chapter 7). (photograph: Margery Wilkins).

All rights reserved. No part of this book may be reproduced, or transmitted, in any form or by any means, electronic, mechanical, photocopying or otherwise, without the prior written permission of the copyright owners.

This book is available direct from Archaeopress or from our website www.archaeopress.com

'*...THE MISSILE IS LAUNCHED WITH SUCH FORCE THAT IT REACHES NOT LESS THAN TWICE THE RANGE OF A SHOT FROM A BOW ...*'

6th century AD writer Procopius,
War against the Goths i.21.17,

describing the bolt launched from the metal frame arch strut catapult.

Greek torsion catapults, which used the energy stored in tensioned and twisted rope springs, were adopted and developed by the Roman army. They were the most powerful missile-projectors of their time in the western world, and had a considerable influence on events there from their invention in the fourth century BC. As bolt-shooters they are still recorded in use by Byzantine armies in the eleventh century AD. No other weapon launching heavy missiles has dominated western warfare for so long.

This research is dedicated to Margery, Ian and Helen, and to my mentors

E.J.W.H., E.W.M., I.A.R., and J.M.C.T.

Contents

List of Figures v

Acknowledgements xi
    The First Edition xi
    The Second and Third Editions xii
    The author xv

Preface xvi
    Archaeological discoveries and recent research xvii

Glossary xxiii

Weights and measures xxv
    Weights and measures xxv
    Vitruvius' list of stone shot weights and corresponding spring-hole diameters (Based on Marsden 1971, 198-199) xxv

Chapter 1. Introduction 1

Chapter 2. Greek origins 4
    The *gastraphetes* ('stomach-bow'): a semi-robotic archer 6
    The introduction of the winch and stand 8
    Torsion-powered artillery 11

Chapter 3. The menace of the new weapon 14
    Sulla's artillery in sieges: the Pompeii 'target' 16
    Caesar's artillery 19

Chapter 4. The bolt-shooter: accuracy, range and effects 22
    Maiden Castle, Dorset 22
    Hod Hill, Dorset. A bolt-shooters' 'target' 24
    The palisaded enclosure and Huts 36a and 37 26
    Which bolt-shooters were used? 26
    Vespasian's battle plan 27
    The evidence of the boltheads 30
    Artillery platforms (*tribunalia*) at Hod Hill 32
    Accuracy 32
    Range 33

Chapter 5. Reconstructing the Roman bolt-shooter 38
    Archaeological evidence for the design of the Roman torsion bolt-shooter 41
    The curious case of the Ampurias frame 41
    The Caminreal iron frame plating 42

    Catapult Washers 43
    The Xanten-Wardt smaller scorpion (*scorpio minor*) 43
    The identity of the Xanten-Wardt catapult 47
    The size formula for bolt-shooters 51
    Constructing the spring-frame 53
    The plating covering the spring-frame 54
    Importing metal plates 54
    The arms 56
    The case and the slider (i.e. the stock) 56
    The trigger assembly 57
    The winch box 57
    The stand and universal joint 57
    The standard bolts 59
    A complete catapult bolt from the Rhine frontier 60
    The Qasr Ibrim foreshaft and iron bodkin heads 63
    Penetration of different types of body armour 64
    The rope springs and the bolt-shooter's performance 64
    Fitting, tensioning and twisting the rope springs 67

Chapter 6. The new design: the metal frame arch strut *cheiroballistra/ manuballista* 69

    Eric Marsden's identification of the machine 72
    The transition from *scorpio* to *manuballista* 75
    The *cheiroballistra's* all-metal spring frame: appraisal of this radically different design 76
    Increase in power 77
    Improved field of view 77
    Faster speed of rewind and consequent increased rate of missile launch 78
    Elimination of ricocheting bolts 79
    The new design of bolt, with aerodynamic and horrific wounding properties 80
    The *Cheiroballistra* manuscripts 82
    Kanones duo (two boards) i.e. Case and Slider 82
    Kleisis (trigger mechanism) 86
    Kambestria (Field-frames) 87
    Pittaria (Pi-brackets) 88
    The Field-frame Rings 90
    Kamarion (Arch) 92
    Klimakion (Ladder) 94
    Kulindroi chalkoi (bronze Cylinders/Washers) and *Kanonia* (Bars) 96
    Konoeide duo (two cone-shaped parts) i.e. Arms 97
    Missing parts 99
    Ease of dismantling and repair 101
    Winch and stand. Double ratchet handles 102

    Tactical positioning of the *manuballista* 105
    The protective metal spring-frame covers 106
    The arm 'sleeves' 108
    Did the arch strut design have inward swinging arms? The evidence of Trajan's Column. 110

Chapter 7. Deciphering the manuscripts: Vitruvius' ballista 113
    The two half-spring frames 115
    The washers 117
    The framework holding the half-springs and the 'inward-swinging arms' theory 117
    The Ladder and the Table 122
    The trigger 122
    Bowstring problems 122
    'Building the Impossible': the BBC *ballista* 125
    The arms 127
    The stand and the angle of elevation 127

Chapter 8. The stone missiles: range and effects 129
    A *ballista* and sling 'target' on the Rhine frontier 131

Chapter 9. Masada AD 73-74 133

Chapter 10. Qasr Ibrim: artillery in defence. Inscribed stone shot 136

Chapter 11. Artillery in action in the field: Arrian's battle plan 145

Chapter 12. Burnswark Roman camps and native hillfort, Dumfriesshire 147
    Evidence for the abandonment of British hillforts 151
    The dating of the site's features and evidence from the Nicholson and Reid excavations 153
    The missile bombardment 158
    The collapse/demolition of the hillfort's rampart 163
    Occupation of the hillfort after the collapse/demolition of the ramparts 164
    Discoveries of sling bullets outside the South Camp 164
    Evidence suggesting that the camps were intended for repeated use 165
    The date(s) of the missile launchings 166
    The Three Brethren as artillery platforms 166
    Training in missile attack and in rapid assault under covering fire 169
    Covering fire 170
    Interpreting the missile bombardment 171
    To summarise 172
    High Rochester *ballistaria*, Northumberland 176

Chapter 13. The last stone-throwers 178
    The one-arm catapult 178
    The traction trebuchet 181

Chapter 14. The Hatra stone-thrower and the inward-swinging arms theory 183
    The Hatra catapult, the Vitruvian *ballista*'s successor?......................... 184
    The inward-swinging arms theory 189
    The Morgan-Feeley Hatra quarter scale models as inward and outward swinging.. 192

Chapter 15. Survival 197

Chapter 16. The Roman achievement 198

Chapter 17. Future search and research 200

Appendix 1. The Roman origin of the mediaeval revolving-nut crossbow release 202

Appendix 2. Review of 2021 TV film on Burnswark Hill, '*Massacre on Hadrian's Wall*' 205
    A programme for the American Smithsonian Channel and Channel Five in Britain. 205

Sources and references to artillery 210
    Accounts of Roman artillery in action 211
    Select bibliography 211
    Museums and sites 216
        Museums 216
        Sites in Britain 219
    Websites and TV programmes 220
    Where the reconstructed catapults described in this book can be seen 221

List of Figures

Len Morgan, the author and Tom Feeley xiii
The dramatic display of the Xanten-Wardt and *manuballista* bolt-shooters at the Carvoran Roman Army Museum was opened by Tom's widow Marian on February 6, 2017 (photograph: Margery Wilkins) xiv
Welcoming the delivery of the giant *ballista* and *onager* at Comlongon Castle. xiv
The BBC one-talent *ballista* team (page 125) is dwarfed by the vast timbers (photograph: Margery Wilkins)..xv
Figure 1 - Technical tests at Piddington Roman villa site xix
Figure 2 - The combined crank handle and handspike from Elenovo, Bulgaria xx
Figure 3 - A line of *ballista* balls and the first complete *scorpio* bolt from the old bed of the Rhine at De Meern, in the Netherlands (Diagram and CT scan: Erik Graafstal)...........xxi
Figure 4 - The spring frame and arms of a reconstructed Xanten-Wardt scorpion bolt-shooter xxiv
Figure 5 - The author's version of Dionysius of Alexandria's *polybolos* 2
Figure 6 - Schramm's fine reconstruction of the polybolos (photo: Dietwulf Baatz)... 3
Figure 7 - Scythian archer. Composite bows 5
Figure 8 - The stomach bow 6
Figure 9 - Testing a *gastraphetes* 7
Figure 10 - Codex P drawing of a gastraphetes (photograph: Wikipedia commons) ... 8
Figure 11 - Catapult winches 10
Figure 12 - Marsden's three-span torsion bolt-shooter (Photograph: Eric Marsden) 12
Figure 13 - The giant BBC ballista, the one talent stone-thrower created for the BBC programme 'Building the Impossible'. 13
Figure 14 - Carthage stone shot 15
Figure 15 - Sulla's artillery: the Pompeii 'target' (Mike Burns. From *JRMES* 14/15, 1-10) ... 17
Figure 16 - Sulla's artillery.. 18
Figure 17 - Alésia 20
Figure 18 - Missiles from Caesar's siege found at Alésia (Mont Auxois in central France) 21
Figure 19 – Boltheads and sling bullets 21
Figure 20 - Artillery victim at Maiden Castle 22
Figure 21 - Marsden's three-span at Maiden Castle: a television first 24
Figure 22 - Plan of boltheads targeting the chieftain's huts and enclosure 25
Figure 23 - Filming at Hod Hill for Channel 5, June 2012 27
Figure 24 - Vespasian's possible plan of attack – crossing the ditches......................... 29
Figure 25 - Reconstruction of an Iron Age hut 29
Figure 26 - Vespasian's plan of attack... 30
Figure 27 - Scorpion and Hod Hill missiles 31
Figure 28 - Close range impact with the *Polybolos* (photograph: the author).............. 33

Figure 29 - Bolt leaving a Xanten-Wardt scorpion 36
Figure 30 - Schramm demonstrating his *ballista* to the Kaiser (photograph: Saalburg Museum) 39
Figure 31 - The Ampurias frame as published 41
Figure 32 - The corrected Ampurias display 42
Figure 33 - The Caminreal (Teruel) iron frame reconstructed 43
Figure 34 - Catapult washers (diagrams after Dietwulf Baatz. Washer photograph: the author) 44
Figure 35 - Vedennius' *scorpio* and the Xanten-Wardt reconstruction 46
Figure 36 - Xanten-Wardt frame 46
Figure 37 - Xanten-Wardt missing parts (photographs: Maarten Dolmans) 48
Figure 38 - Operating Xanten-Wardt scorpions 50
Figure 39 - The Xanten-Wardt Report's box support (after Schalles 2010, 175 Figure 5) 50
Figure 40 - Reconstruction of the Xanten-Wardt scorpion 51
Figure 41 - Carlisle catapult find 52
Figure 42 - Frame wedge system on a *scorpio maior* 55
Figure 43 - Scorpion case, slider and winch (photographs: the author) 57
Figure 44 - Catapult on the Altar of Zeus, Pergamon. Scorpion curved arms (photograph: Margery Wilkins) 58
Figure 45 - The Cremona battle-shield 59
Figure 46 - Scorpion stand 60
Figure 47 - The complete number 211 bolt as conserved 61
Figure 48 – Bolt 211 (top): narrow glue or paint bands at the rear end. Bolt 416 (bottom) with incomplete shaft 61
Figure 49 - Two scorpion reconstructions 62
Figure 50 - Qasr Ibrim foreshaft and bolthead 63
Figure 51 - Bolt damage to lorica segmentata 64
Figure 52 - Bolt strike on chainmail and helmet 64
Figure 53 - Artillery boltheads 66
Figure 54 - William Newton's version of the rope tensioning frame described by Vitruvius and Heron 68
Figure 55 - Catapult scene on Trajan's Column, Rome (cast in the Victoria and Albert Museum, London) 69
Figure 56 - Victor Prou's *cheiroballistra* (from Prou 1877 Figures 47 and 16) ... 71
Figure 57 - Trajan's Column catapults 71
Figure 58 - Eric Marsden's *cheiroballistra* 72
Figure 59 - Two *carroballistae* on Trajan's Column 73
Figure 60 - The author's first interpretation of the *Cheiroballistra* manuscript, built in 1991 74
Figure 61 - Aiming the three-span scorpio and the *cheiroballistra* (photographs: the author and Margery Wilkins) 78
Figure 62 - Misfire (still frame from video: the author) 79

Figure 63 - Dura Europos bolt replica 80
Figure 64 - The horrific strike effect of this bolt on ballistic gel simulating human or animal flesh 81
Figure 65 - Slow-motion film of a *cheiroballistra* bolt penetrating a large melon 82
Figure 66a - The Cheiroballistra manuscript 83
Figure 66b - The Cheiroballistra manuscript 84
Figure 67 - Operating the *cheiroballistra* 85
Figure 68 - *Cheiroballistra* trigger mechanism 86
Figure 69 - Field-frame replicas and Gornea example 87
Figure 70 - Codex M's Field-frames and Heron's ballista diagram 88
Figure 71 - Junction of the Ladder tenons and the Field-frames (models and photographs: the author) 89
Figure 72 – Lyon and Orşova Field-frames 91
Figure 73 - The Elenovo *Kambestrion* (photographs: Kayumov and Minchev) 91
Figure 74 - *Kambestria* from Elenovo and Sala (Morocco) 92
Figure 75 - Orşova iron Arch strut (photograph: Dietwulf Baatz) 93
Figure 76 - The *Cheiroballistra* Arch 93
Figure 77 - Codex P Arch and Ladder diagram 94
Figure 78 - Reconstruction of the main frame. The Ladder 95
Figure 79 - Codices M and V diagrams of the Arms (diagrams after Wescher) 98
Figure 80 - The Elenovo arm (photograph: Kayumov and Minchev) 98
Figure 81 - Tom Feeley's version of the alternative wood encased Arms (photograph: Tom Feeley) 99
Figure 82 - The position of the Ladder on the Field-frames. See text below (photograph: the author) 99
Figure 83 - Left hand end of the reconstructed *cheiroballistra* 101
Figure 84 – The author's version of the *cheiroballistra* frame parts 102
Figure 85 - The Cupid Gem 103
Figure 86 - Reconstruction of the *cheiroballistra* 103
Figure 87 - The Elenovo crank handle and handspike (illustration from Kayumov and Minchev 2010) 104
Figure 88 - Reconstruction of the Elenovo crank handle-handspike (photographs: Tom Feeley) 105
Figure 89 - Trajan's Column: *ballista* on cart (photograph of the Column: the author) 106
Figure 90 - The Lyon size of arch strut catapult: Tom Feeley's version 107
Figure 91 - Adding extra torsion to Len Morgan's Lyon size catapult (photograph: Margery Wilkins) 107
Figure 92 - The Arm 'sleeves' 108
Figure 93 - Two reconstructions of the *cheiroballistra* (photographs: the author) 109
Figure 94 - Tom Feeley's framework. Impact of bolt on *lorica segmentata* 109
Figure 95 - Penetration tests of bolts (photographs: the author) 110
Figure 96 - Trajan's Column: projections on the outside of the field-frame covers (photographs: the author) 111

Figure 97 - Trajan's Column: forward projecting beams – the sliders 112
Figure 98 - Len Morgan's two *ballista* reconstructions 113
Figure 99 - *Ballista* half spring-frame 116
Figure 100 - Heron's rhombus diagram 116
Figure 101 - Vitruvius' plated washers 117
Figure 102 - Heron's *ballista* diagram 118
Figure 103 - *Ballista* model made for Eric Marsden (photograph: the author) 119
Figure 104 - Crossbeams 119
Figure 105 - Diagram of the *ballista* framework 120
Figure 106 - The crossbars and the table 122
Figure 107 - The author's model *ballista* 123
Figure 108 - Trigger and mounting plate 124
Figure 109 - The bowstring belt 124
Figure 110 - Cross-section of table, ladder and slider 125
Figure 111 - The BBC one talent Vitruvian *ballista* (photograph: the author) 126
Figure 112 - The Morgan-Wilkins two *librae* catapult in action 128
Figure 113 - Cutting a one talent stone ball 129
Figure 114 - Philon's defence against the one talent *ballista* 131
Figure 115 - *Ballista* stones and sling shot in a long line on the former river bed close to the De Meern fort. 131
Figure 116 - Stone missiles at the Masada Exhibition (photograph: Eric Marsden) 133
Figure 117 - Qasr Ibrim before the Aswan Dam 136
Figure 118 - Map of the Nile. Airphoto of the site 138
Figure 119 - *Ballista* balls in South Rampart Street (photograph: McDonald Institute Cambridge) 138
Figure 120 - Ink writing on *ballista* balls 139
Figure 121 - Hans Barnard's size analysis. The seven British Museum balls 140
Figure 122 - Clusters of *ballista* balls on the West Rampart 141
Figure 123 - The Kandaxe ball and Message for Hitler on a WW2 shell 142
Figure 124 - Qasr Ibrim tanged boltheads. 143
Figure 125 - Burnswark Hill from the SE, from close to the Eaglesfield crossing of the M74 (photograph: the author) 147
Figure 126 - Burnswark hillfort and the Roman South Camp (Reproduced with the permission of Cambridge University Collection of Aerial Photography © Copyright reserved.) 147
Figure 127 - Burnswark hillfort and Roman camps, Dumfries and Galloway (from Jobey 1977-8 Figure 1) 149
Figure 128 - The Roman camps and hillfort from the north east (photograph: © Crown Copyright: HES) 150
Figure 129. *Ballista* ball from the 2015 excavation Trench 2 152
Figure 130 - The 2016 Trench 4 cut into the back of the South Camp's north west rampart (photograph: Andrew Nicholson) 154

Figure 131 - The southern Roman camp, Burnswark, looking south (photograph: © Crown Copyright: HES) 156
Figure 132 - Burnswark sling bullets, arrowheads and a Trajan's Column stone thrower and slinger 157
Figure 133 - Burnswark missiles (photograph: John Reid) 158
Figure 134 - The Three Brethren and the 2016 trenches behind the rampart 159
Figure 135 - Groups of lead sling bullets found in Trench 5 of the 2016 excavations and in the North Camp 160
Figure 136 - Excavation trenches 1-5 and the distribution of lead signals from 2013 to 2017 161
Figure 137 - Carol van Driel-Murray at the centre one of the Three Brethren catapult mounds 162
Figure 138 - Carol standing in the South Camp's deep recut ditch west of the centre gateway 162
Figure 139. The 2015 excavations on Burnswark Hill 163
Figure 140 - Trajan's Column Scene LXVI. An arch strut bolt-shooter on a log emplacement 167
Figure 141 - 'Troops' passing through the centre gateway, thirteen abreast. (photograph: John Reid) 169
Figure 142 - The BBC one talent ballista launching the final ball (photograph: Margery Wilkins) 170
Figure 143 - A two *librae* Vitruvian *ballista* being assembled by a team of legionaries (photograph: the Roman Military Research Society) 175
Figure 144. High Rochester *ballistaria* inscription (after Richmond). 176
Figure 145 - Ms diagram of a one-arm stone-thrower on a siege tower 178
Figure 146 - Two modern reconstructions of the onager 180
Figure 147 - The Morgan-Wilkins onager 181
Figure 148 - The Traction trebuchet (Wikipedia Commons) 182
Figure 149 - Excavation photo of the Hatra catapult (photograph: W.I. Al-Salihi, Directorate-General of Antiquities, Baghdad.) 183
Figure 150 - Mosul Museum's strange reconstruction of the Hatra catapult (photograph: Mary Desbruslais) 185
Figure 151 - Baatz's diagram of the Hatra frame 186
Figure 152 - The author's suggested spring-frame design of the Hatra catapult. The bronze washers. (photographs: Diewulf Baatz) 187
Figure 153 - The Morgan-Wilkins outward swinging Hatra catapult 188
Figure 154 - Martin McAree's graphs of the relative performances of inswinger and outswinger 191
Figure 155 - Mike Lewis' models 192
Figure 156 - Hatra inswinging designs: (left) Aitor Iriarte's two versions of the Hatra catapult. (right) Mike Lewis' inswinging *cheiroballistra*. 192
Figure 157 - Tom Feeley's quarter-scale inward swinging Hatra catapult 193

Figure 158 - Tom Feeley's inswinging version in three positions (photographs: Mark Hatch) 194
Figure 159 - A battery of twelve bolt-shooting and one stone-throwing catapults, June 2013 195
Figure 160 - The fragmentary Roman hunting relief from St Marcel 202
Figure 161 - The Solignac hunting relief 203

Acknowledgements

The First Edition

The untimely death of Dr Eric Marsden in his late forties robbed the world of Greek and Roman studies of a brilliant mind. I have tried to keep up the momentum of his research, attempting fresh reconstructions based firmly on the manuscript evidence and archaeological finds. All those working in this field owe an immense debt to Marsden and to a long line of scholars and enthusiasts, outstanding among whom were the experimental archaeologist Erwin Schramm, and the text editor Carle Wescher. Professor Dr Dietwulf Baatz has long maintained an eagle-eyed watch on the discoveries of catapult parts, publishing them with exemplary clarity. I am very grateful to him for advice and for allowing me to reproduce some of his drawings and photographs. It is a great delight to acknowledge the considerable help and constant encouragement of Mrs Margaret Marsden and of Dr John Marsden, Eric's son. I have benefited from correspondence or conversations with many people, notably Don Acklam, Lindsay Allason-Jones, the late John Anstee, Caroline Baillie, Hans Barnard, Mike Bishop, Duncan Campbell, Tom Feeley, Rex Harpham, Mark Hassall, Aitor Iriarte, Simon James, Lawrence Keppie, Carole Kirsopp, Gordon Macdonald, Christos Makrypoulias, Steve Ralphs, Pam Rose, David Sim, Digby Stevenson, and Carol van Driel Murray. The credit for undertaking the daunting challenge of building the one-talent *ballista* goes to producers Helen Thomas, George Williams and their BBC colleagues. I am grateful to Adam Hart-Davis and producer Martin Mortimore for challenging me to reconstruct Dionysius' repeating bolt-shooter. My colleagues in the Roman Military Research Society and Paul and Christine Jones have helped with the field tests. Permission to use illustrations has been given by The Society of Antiquaries of London, Cambridge University Library of Air Photographs, Simon James, Len Morgan, Tom Feeley, the Ermine Street Guard, Dumfries Museum, Saalburg Museum, Staatliche Museum of Berlin. The Victoria and Albert Museum, London kindly allowed photography of the casts of Trajan's Column. The staff of the British Museum were extremely helpful in allowing photography and weighing of the Qasr Ibrim stone shot.

It will be obvious that I owe a large and continuing debt to Len Morgan for realising my ideas so brilliantly. The superb quality of his reconstructions of the *scorpio* and *cheiroballistra* can be seen in the following pages and at the displays of the Roman Military Research Society. Sir Ian Richmond started my interest in the subject with the loan of Schneider's book on Schramm's reconstructions.

The Second and Third Editions

Several of the above have continued to contribute very valuable help or advice, most notably my engineer friends and collaborators Len Morgan and Tom Feeley, whose catapult construction has proceeded apace. I continue to owe an enormous debt to them. Unlike the general technicians who in the past built machines for previous researchers such as de Reffye, Dufour, Schramm and Marsden, Len and Tom already had an extensive knowledge of Roman technology and long experience of reconstructing Roman arms and armour before they linked into my research. Len has that special skill as a pattern-maker, and his superb replicas and reconstructions of armour and weapons are found in museums and displays all over the Roman Empire.

Very sadly, Tom died of cancer on 30 November 2015. The photographs of the machines which he made in conjunction with Len Morgan, confirm the staggering quality of his metalwork in particular. After a long and distinguished career in the oil industry, he had devoted many years to a study of the equipment of the Roman army, reproducing armour, weapons, furniture and so on with the highest skill and authenticity.

Len Morgan and I met in 1993, and displayed our first reconstruction of the *cheiroballistra* in 1995 at Leiden University. Over subsequent years we worked on the Vitruvian bolt-shooter and stone-thrower, based on my revised editions of Vitruvius' text, the Spanish frames and the Cremona shield. In 2004 Tom joined our programme of catapult reconstructions. His tremendous drive, enthusiasm and determination to work through difficult challenges enabled him to solve numerous problems, such as how to create the 'cans' covering the spring frames of the *cheiroballistra,* including their projecting 'sleeves' (Figure 92), the soldering technique required to create the decorative edging of the Xanten-Wardt catapult's bronze plates (Figure 40), and how to construct the eight-sided housing of the Elenovo crank-handles (Figures 2 and 88). He was the first person to solve the latter problem, with Len's help, mounting a pair of these crank handles on the winch of his *carroballista* size arch strut catapult, and drawing important conclusions about how this design of handle speeded up loading time and therefore the rate of missile launch.

David Breeze continues to give me much valued support and is responsible for this Third Edition being published by Archaeopress. Our son Ian has always maintained a fine computer set up for me and has given vital help in preparing this edition for the publisher. My wife Margery has shown endless patience in allowing me to give priority to the Roman world, rather than to that of twenty-first century house maintenance. This book is for her and would never have happened without her support.

I recall with gratitude Karen Carruthers and Janet Watson's early help in translating Baatz's technical German. I also owe thanks to Mark Hatch and my colleagues in the Roman Military Research Society, Richard Abdy, Andrew and Barbara Birley, Hans Barnard, Mike Bishop, Dot Boughton, Salvador Busquets, The Dorset County Museum, Duncan Campbell, Mary Desbruslais, Maarten Dolmans, Carol van Driel-Murray, the Elder family, the Ermine Street Guard, Roy Friendship-Taylor, Adam Hart-Davis, Erik Graafstal, Maureen Guirdham, Bill Hanson, Rebecca Jones, Ildar Kayumov and Alexander Minchev, Ralph Jackson, Kay Kingstone, Carolina Rangel de Lima, Martin McAree, the Marsden family, Andrew Nicholson, Tim Padley, Graham Piddock, John Reid, Lucy Romeril, Pamela Rose, David Sim, John Smith and Marcus Brittain, Digby Stevenson, David Thomson and Teresa Church, Derek Welsby, Alexander Zimmerman and Jo Kempkens.

Len Morgan, the author and Tom Feeley
Our many years of friendship and enthusiastic cooperation are responsible for the catapult reconstructions recorded in this book. Tom left his Xanten-Wardt and *manuballista* bolt-shooters to the Carvoran Roman Army Museum on Hadrian's Wall.

The dramatic display of the Xanten-Wardt and manuballista *bolt-shooters at the Carvoran Roman Army Museum was opened by Tom's widow Marian on February 6, 2017 (photograph: Margery Wilkins)*

Welcoming the delivery of the giant ballista *and* onager *at Comlongon Castle.*
The future custodians of our artillery, Teresa Church and Professor David Thomson, accepting delivery of the components of the ballista *and the* onager.

The BBC one-talent ballista team (page 125) is dwarfed by the vast timbers (photograph: Margery Wilkins).

The author

Alan Wilkins studied Classics at Lancaster Royal Grammar School and read the subject at Emmanuel College, Cambridge, specialising in ancient history and archaeology under Professors Jocelyn Toynbee and A. G. Woodhead. He spent several years excavating on Roman military and civilian sites in Britain, and was a field assistant to Sir Ian Richmond for 17 years. He lectured on Greek and Roman Civilisation for Liverpool University's Extra-Mural Department, and was one of the pioneers of the JACT evidence-based teaching of Greek and Roman history. After 30 years teaching Classics at Woodbridge School, Merchant Taylors School, Crosby and Annan Academy, he turned to the subject of Greek and Roman artillery, following the tragic early death of his friend Dr Eric Marsden. He has attempted to maintain the momentum of Eric's research into the subject. He is a Fellow of the Society of Antiquaries of London.

Preface

This Third Edition has been extensively revised and updated, with more illustrations and new sections on the exciting finds of the first complete catapult shaft and bolthead and a line of *ballista* balls from the old bed of the Rhine at Utrecht (Figure 3), the film of the horrific wounding properties of a *cheiroballistra* bolt shot into ballistic gel Figure 64), and the two reliefs in France of Roman hunting crossbows which prove that they are the ancestors of mediaeval crossbows (Figures 160 and 161). I have completely rewritten the section on the Burnswark camps, offering decisive arguments for the site as the only known Roman army practice area.

The second edition covered the latest discoveries, the author's revised editions of the texts of the Greek and Roman engineers, and full-size reconstructions of the bolt-shooting and stone-throwing catapults. Nearly all of the latter have been produced in collaboration with engineers Len Morgan and Tom Feeley, and are based firmly on the ancient texts, the archaeological finds, and evidence from Roman relief sculpture.

After tracing the Greek origins of torsion-powered catapults, this book describes the machines used from the time of Sulla and Caesar onwards, their dominance in the warfare of the western world for over a thousand years, and their importance in the history of technology. The exciting find of a nearly complete catapult frame at Xanten-Wardt in Germany allows the Roman army's bolt-shooter to be reconstructed with a very high degree of accuracy. This type is identified as the one used extensively in the invasion of Britain; as a result, a modified scenario is suggested for Vespasian's attack on the hill fort of Hod Hill, Dorset.

The author's 2007 revised edition of the Latin text of Vitruvius' description of the stone-throwing *ballista* replaces the 1917 unsatisfactory and misleading version. From it Len Morgan has built two impressive full-size *ballistae* which have proved successful in action. From Egypt a vivid picture has emerged of a small garrison of legionaries, isolated in enemy territory down the Nile, toiling to cut stone ammunition for these machines. The unique ink inscriptions on several of the stone balls tell the story of the rivalry between five centurions. At the siege of Masada in Israel large stone-throwers were a key part of the attack on the Jewish resistance fighters in Herod's spectacular fortress, bringing the war against the Jews to an end.

The only surviving Roman battle plan by a Roman general, Arrian, shows how artillery was deployed in the field to halt an attack in Cappadocia (East Turkey) by a large force of the feared Sarmatian mounted spearmen. The technique was to put down such an '*indescribable volume of missiles*' from the catapults, in combination with archers, slingers and spearmen, that the charging Sarmatians would never reach the legionary infantry line.

Technical tests, conducted in cooperation with David Sim, demonstrate the penetration power of bolt-shooters. The range and accuracy achieved by Roman catapults is discussed. All of this accounts for the superiority of torsion catapults over conventional weapons, and their ability to ward off most threats to the peace of the Roman world.

To attempt to settle the controversy as to whether the catapults with wide frames had inward swinging arms, my two colleagues constructed quarter-scale models of the Hatra catapult, one as an inswinger, the other as a conventional outswinger. The fascinating results of shooting and other technical tests are given towards the end of the book.

Archaeological discoveries and recent research

Major archaeological discoveries and many developments in research on Roman artillery have been made since the Shire publication of my *Roman Artillery* in 2003.

The most important new information is the detailed publication in 2010 of the nearly complete main frame of a bolt-shooting catapult of the mid first century AD, found at Xanten-Wardt in north west Germany. This exciting find has added enormously to our understanding of these machines, because not only have the metal plating and washers survived, but for the first time the wood of a catapult frame and the front end of the wooden case and slider have been preserved. This means that it is no longer necessary to estimate the possible details of this standard Roman bolt-shooting catapult. German and Dutch experts responsible for conserving and reconstructing the frame kindly sent a full set of one-to-one plans to enable my highly skilled engineer colleagues Len Morgan and Tom Feeley to make millimetre-accurate copies. I have prepared one of these for the British Museum's 2024 Exhibition *Legion: Life in the Roman Army*.

From this discovery we have been able to confirm that two pieces of iron-bound ash, found in a Roman workshop during the Carlisle Millenium Excavations, are parts under construction for the same size and type of bolt-shooter as the Xanten-Wardt. This is the first hard evidence that standardised catapult plans were distributed to Roman army workshops throughout the Empire. The Carlisle objects are also the first parts of the framework of a catapult to be identified in Britain.

The fact that the rope springs of the Xanten-Wardt *scorpio* have the same height as those of the *cheiroballistra,* and have the same spring diameter, 45mm, as Eric Marsden's corrected reading for the *Cheiroballistra* manuscript, producing a spring volume of 0.42litre, suggests that the designers of the *cheiroballistra* adopted the Xanten-Wardt springs, and throws further doubt on Aitor Iriarte's non-winched version with 25.6mm diameter springs and a 0.13 litre spring volume.

During investigations into the sizes of catapult boltheads found on sites like Hod Hill and Vindolanda, I have found that the majority have an internal socket diameter of 10 – 12mm. This size is too small for a three-span, but exactly right for the Xanten-Wardt bolt-shooter, suggesting that this highly portable version was a very common field machine, probably producing a very effective battlefield performance from an economical amount of sinew rope. The Xanten-Wardt appears to be the Lee-Enfield or Mauser of its day.

Pamela Rose, Hans Barnard and I have published a detailed account of the brief Roman military occupation of Qasr Ibrim on the Nile, described by Strabo. Pamela Rose has described the excavations of the defences and pinpointed the positions of caches of *ballista* balls. The many hundreds of them surviving from excavations on the site have been weighed and analysed in great detail in a *tour de force* by Hans Barnard, the first large cache of Roman artillery balls of Imperial date to be fully published at the time. My decipherment of the unique ink inscriptions on 38 of the balls has shown that the garrison was under the command of five centurions, four of whom had assumed the famous names Pompey, Julius, Antony and Octavius. The inscriptions and the differing weights of the stone shot allow estimates of the sizes of stone-throwing *ballistae* used and other technical details of the machines.

The *Proceedings of the VII Roman Military Equipment Conference, Zagreb* contains the important report by Ildar Kayumov and Alexander Minchev on the military equipment ploughed up in 1962 in Southern Bulgaria. Three iron items are related to artillery, a *kambestrion* field-frame, a crank handle-handspike and a possible arm.

My 2007 edition of the Latin text by Vitruvius describing the stone-throwing *ballista* is the first radically revised edition to appear since the unsatisfactory and misleading version by Schramm and Diels in 1917. It confirms Eric Marsden's discovery that Schramm included '*a number of items that would seem never to have been in Vitruvius' original*', and disproves a recent claim that the Latin text is '*badly mutilated*' and an invalid source for understanding the machine. My large model is based on this revised edition and the descriptions of Philon and Heron. Len Morgan has used it to construct two magnificent full-size versions of a two *librae ballista*.

The late Tom Feeley produced not only the first reconstruction of the protective covers on the arch strut catapults depicted on all the Trajan's Column machines, but also, along with Len Morgan, the first reconstruction of the Lyon size of arch strut bolt-shooter, which was probably the size used as *carroballistae*. All Tom and Len's *carroballistae* need are a cart and two mules!

We have collaborated with David Sim in live-firing field tests to calculate the actual penetration power and damage effects of Roman bolt-shooting catapults. The results have already modified some claims by writers on this subject.

Figure 1 - Technical tests at Piddington Roman villa site
A battery of 12 bolt-shooters and at the far end the Morgan-Wilkins stone-thrower. The bolts were shot at various types of replica armour. **Inset**: David Sim recording the hits on his group of hand-forged replica shield bosses.

David Sim has made the exciting discovery that the Roman army was using <u>rolled</u> iron plates of uniform thickness (variation of no more than 0.1mm), possibly importing them from India. I believe that there is a further possibility that the technique or even the supply of rolled plates actually originated from China along the Silk Road. These plates were used to produce the boltheads for catapults (Sim 1995, 1-3), and possibly for the plating on catapult frames. The first rolling mills are customarily stated to have appeared in Europe during the 18th century.

The arguments for the existence of catapults with inswinging arms have tended to become a theoretical exercise *in vacuo*, losing sight of the artillerymen's battlefield priorities. Of course, when it is proposed that inswinging arms would have an arm arc of 100+° for the arch strut bolt-shooters, as opposed to the 70° of an outswinging version, the conclusion does indeed appear to be obvious, a 'no brainer' – inswingers should produce more power, which is, after all, the aim of catapult construction as defined by Philon and Heron. Practical tests are one valid approach, and Len Morgan and Tom Feeley completed quarter scale versions of two opposing interpretations of the Hatra catapult, as an inswinger and outswinger. The results, incorporating graphs of technical analysis, are described in Chapter 14.

Even before these tests, I pointed out that the additional time taken to wind back the inswinging arms through the extra 30° or so would have markedly reduced the rate of missile launch: no Roman artilleryman faced with a rapidly approaching enemy would thank you for that. Hadrian himself emphasises that the priority for archers was to practise to achieve faster speeds of shooting, (page 196).

One problem in particular, the shortness of the inswinger's arms, means that as short levers they make the winding back of the extra 30° much harder and slower. The existence of two improved, faster rewind winch systems, the Cupid Gem twin handle 'seesaw' action and the Bulgarian crank handle-handspike, confirms that the Romans were intent on speeding up rewind to increase the volume of missiles. As to *palintone* catapults, my establishment of a Latin text of Vitruvius free of the Schramm-Diels '*mutilations*', and Len Morgan's resulting construction of two full-size Vitruvian *ballistae* have confirmed that Vitruvius' and Heron's *palintone* stone-throwers had outward swinging arms.

In his very important scholarly *Hesperia* article (Campbell 2011) Duncan Campbell warns against the damage to artillery research done by 'blue-sky' thinking (in the pejorative meaning of that phrase). '*Schneider's bold hypothesis, that the text labelled with the name of a catapult (for what else could a cheiroballistra be?) was, in fact, no such thing, effectively derailed the study of the iron-framed ballista and took it down a blind alleyway, where it remained for 60 years... It was only after the text was rescued by Eric Marsden that it was again taken seriously as a description of a catapult. If we are to maintain the rigor of our discipline we must be careful to rein in the kind of 'blue-sky' thinking that Schneider freely employed, or at least subject it to careful scrutiny.*' (Campbell 2011, 678). Similar damage has been done over the last 14 years by Aitor Iriarte's Procrustean solution of the *Cheiroballistra* text: his dogged determination to retain the 25mm rope spring diameter of the ms reading has entailed forcing details to fit. He was, to his great

Figure 2 - The combined crank handle and handspike from Elenovo, Bulgaria
The 1962 find was identified by Kayumov and Minchev amongst a group of Roman military artefacts which include a catapult spring-frame and possible arm. (Photo: Kayumov and Minchev)

credit, honest enough to admit twice (Iriarte 2000, 58 and 62) that, 'As Wilkins remarks, it is impossible to pass the tenons of the 'ladder', which are 3 d<actyls> apart, through the pitaria [pi-brackets] in the kambestria [field-frames]. I have no defence at this point...' The actual ms measurements and the resulting relationship of tenons to pi-brackets are

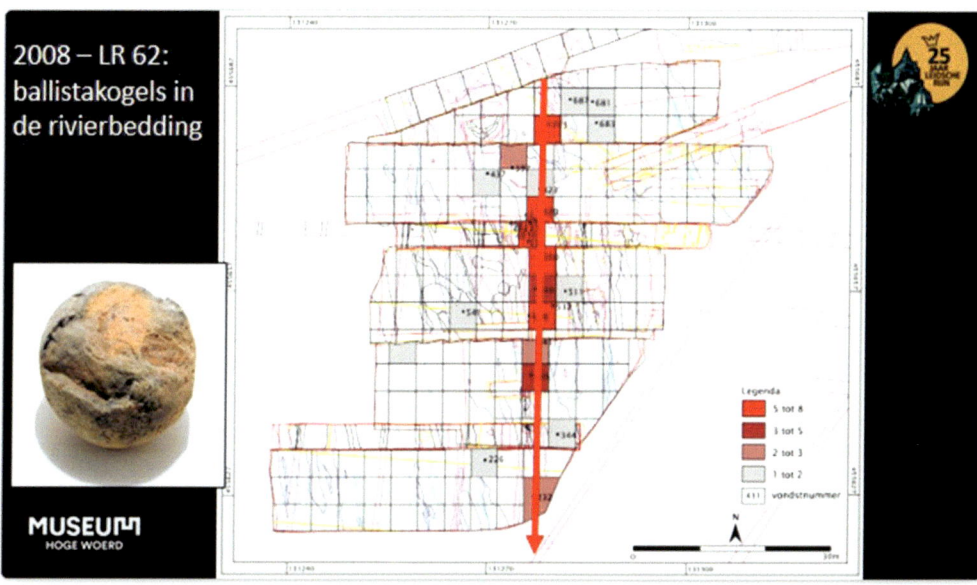

Figure 3 - A line of ballista *balls and the first complete* scorpio *bolt from the old bed of the Rhine at De Meern, in the Netherlands (Diagram and CT scan: Erik Graafstal)*
See Chapter 5, page 60 and Figures 47 and 48 for the Rhine bolt, and Chapter 8, page 131 and Figure 115 for the line of *ballista* balls.

discussed in Section 7. Unfortunately, he then ignored this impossibility, '*conjured up folds... to set the tenons apart enough*' and passed his tenons through the pi-brackets. As a consequence, all other reconstructions of which I am aware have followed his lead and forced the tenons through the pi-brackets; they are, unfortunately, invalid as interpretations of the *Cheiroballistra* ms. Other examples of desperate 'blue-sky' thinking are 1. Kayumov and Minchev's explanation (2010, 339) that the beams protruding from the frames of the Trajan's Column catapults cannot be sliders [which are moved forward at the start of the loading procedure in order to lock the trigger on the bowstring] because '*a slider was a quite fragile element...A Roman artilleryman would hardly leave his engine in such a vulnerable state out of fear of accidental damage.*' 2. Iriarte's belief (2000, 70) that the round 'cans' covering the field-frames on all the Trajan's Column machines bear an '*enormous*' resemblance to the washers, field-frames and spring-cord and are stylized representations of them, leads him to explain the absence of the lower washers on the Column scenes by saying '*Perhaps the chief sculptor thought that the upside down washers - such components would, 'no doubt', fall by their own weight - were a mistake in the cartoons and , subsequently, eliminated them.*'

In conclusion, developments such as the publications of the Xanten-Wardt frame and the 1962 Bulgarian finds, plus the Morgan-Wilkins work on Vitruvius' palintone *ballista*, and the Feeley-Morgan tests on inswinging and outswinging catapults, invite revised, constructive thoughts from all interested scholars. We look forward to reading them.

Glossary

Antonine: dating to the reigns of Antoninus Pius, Marcus Aurelius and Commodus, AD 138 - 192.

ballista (plural ballistae): stone-throwing catapult (but by AD 100 used for a bolt-shooter).

ballistarium: artillery emplacement (Figure 144).

ballistarius (plural ballistarii): artilleryman, artillery mechanic.

carroballista (plural carroballistae): bolt-shooter mounted on a cart (see Figure 59).

case and slider: the two components of the stock of a catapult (see Figure 4 below).

catapulta (plural catapultae): bolt-shooting catapult.

cheiroballistra / manuballista: the revised design of bolt-shooter, introduced by AD 101-2.

cylinders: see washers.

euthytone: 'stretched straight', describing a bolt-shooter whose springs are mounted in a frame with stanchions (upright beams) which, when viewed from above, appear to be in a straight line (see Figure 40).

Hadrianic: dating to the reign of Hadrian AD 117 – 138.

hole-carriers: the horizontal beams forming the top and bottom of the spring-frame, and pierced with holes for the rope springs, their washers and the washer pins.

palintone: 'stretched back', describing a catapult whose springs are mounted in a frame with stanchions which, when viewed from above, are visibly staggered and not in a straight line, allowing the arms to travel in a greater arc than in a euthytone catapult. (See Figure 105).

scorpio (plural scorpiones): a scorpion, either an alternative word for *catapulta* (q.v.), or a specific size of small bolt-shooter – the Larger or the Smaller Scorpion.

spring-frame: the frame containing the rope 'springs' (see Figure 4).

stanchions, centre- and side-: the upright beams of the spring-frame

three-span: shooting a bolt three times a hand span, i.e. 27 Roman inches (69 cm) long. A popular size, combining reasonable portability with great power.

torsion catapult: so called because the springs of sinew rope are first stretched very tightly around the washer bars, and then the action of winding back the arms adds torsion (twisting) to them, creating an enormous amount of stored energy in the rope-spring. If the rope-spring loses its tension, further torsion can be added by twisting the washers with a spanner.

two-cubit: shooting a bolt two cubits, i.e. 36 Roman inches (92 cm) long.

universal joint: the flexible link between the stand and the stock of the catapult (see Figure 4).

washers/cylinders: the four cylinders and their bars through which the two skeins of sinew rope are stretched.

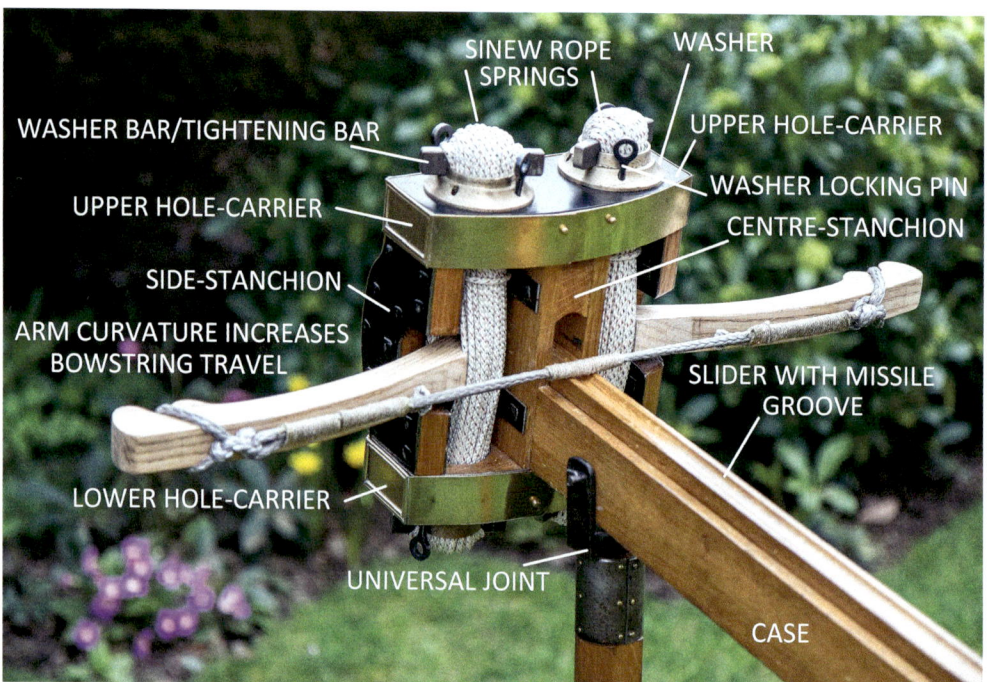

Figure 4 - The spring frame and arms of a reconstructed Xanten-Wardt scorpion bolt-shooter

Weights and measures

Weights and measures

one Greek *mina* = 0.436kg

one Greek *dactyl* = 19.268mm

one *talent* = 26.196kg

one cubit = 46.2cm

one Roman *libra* (pound) = 0.327kg

one Roman *uncia* (inch) = 24.6mm

Vitruvius' list of stone shot weights and corresponding spring-hole diameters (Based on Marsden 1971, 198-199)

Weight of shot		Spring-hole diameter	
librae	kg	*unciae*	cm
2	0.655	5	12.3
4	1.31	6	14.76
6	1.96	7	17.22
10	3.27	8	19.68
20	6.55	10	24.6
40	13.1	12¾	31.3
60	19.65	14⅛	34.75
80	26.2	15	36.9
120	39.3	17½	43
160	52.4	20	49.2
180	58.95	21	51.6
200	65.5	22	54.1
240	78.6	23	56.5
360	117.9	24	59

Chapter 1

Introduction

'The object of catapult construction is to project the missile over a great distance and strike a hard blow at a given target.'
(Heron of Alexandria, *Belopoiika* 74)

This book is an introduction to the catapults of the Roman army, their influence on the history of the western world, and their importance in the history of technology. There is a vast amount of material on this subject in the form of written sources, archaeological evidence and modern attempts at reconstruction. After examining the Greek origins of artillery, this account concentrates on the catapults used by Roman armies from the time of Caesar to the late Roman Empire.

Because only a small fraction of the written records of the Greek and Roman world has survived, the importance of the remaining artillery texts is far greater than their face value as descriptions of machines of war. There are some technical works on other subjects, but artillery is the one field of Greek and Roman applied technology for which there is extensive, detailed evidence; this is because of the survival of several technical manuals by specialist artillery engineers, in particular the Greeks Philon of Byzantium and Heron of Alexandria, and the Roman Vitruvius (see Sources). In addition to recording lists of parts and dimensions, they provide details of the development of catapult design, the workshop methods of engineers, their solutions to problems, and their willingness to push forward into uncharted areas of mathematics, physics and metallurgy. There are also the numerous passages describing artillery in action that are found in Greek and Roman historians, many of whom, such as Caesar, Arrian and Ammianus, were the very army officers who deployed these machines.

The *polybolos* is a *skorpidion* ('small scorpion') torsion bolt-shooter described in detail by Philon of Byzantium (Marsden 1971, 147-151 for text and translation). It is a semi-automatic repeating catapult which may date from *c.* 300 BC, and is an impressive compendium of Greek technical inventions, starting with double chain drive running over pentagonal sprockets to move the slider forwards and backwards, and a cam mechanism translating this linear movement into rotary motion to turn the long roller which feeds the bolts one at a time by gravity into the groove on the slider. Both these devices are their first known appearances in the field of western technology. Trip mechanisms cause the trigger to capture and eventually release the bowstring. The timing of all the mechanisms is exquisite. The special tripod base has a universal

Figure 5 - The author's version of Dionysius of Alexandria's polybolos
Loading the gravity-fed magazine and pulling back the slider and bowstring of the author's reconstruction of Dionysius of Alexandria's *polybolos* ('multishooter'). (Photo: Geoffrey Hircombe)

joint to allow the machine to be swivelled, and a back stay with a prop that moves in a horizontal sliding groove, enabling the catapult's elevation to be adjusted and locked, as Philon states,'by a small lathe-cut (= screw-threaded?) bar or wedge'. Thus, the machine could be locked onto a target, as with the tripod mounts of modern light or medium machine guns, and it only required the winch handles to be turned continually to launch the bolts that had been loaded into the gravity-feed magazine. Philon says that the design had not found a noteworthy use, and there is no evidence of its use by the Roman army, although it would have been surprising if Roman engineers had not evaluated its potential. It was probably too complicated, and the pins connecting each link of the chains would have had to stand up to enormous stresses to avoid these becoming the catapult's Achilles' heel.

Greek artillery engineers were the first to make use of cube root equations; they employed devices such as straight and rotary ratchets, winches, multiple pulley systems, horizontal sliding dovetails, flexible universal joint bearings, and the enormous energy stored in stretched and twisted sinew-rope springs. Some of these devices were in everyday use for non-military purposes, but without the artillery manuals they would be unrecorded or imperfectly understood. The Greek engineers' bold ideas for catapults powered by bronze springs or compressed air pistons outstripped the limitations of contemporary technology.

Their torsion-powered catapults, which used the energy stored in tensioned and twisted rope springs, were adopted and developed by the Roman army. They were the most powerful missile-projectors of their time in the western world, and had a considerable influence on events there from their invention in the fourth century BC. As bolt-shooters they are still recorded in use by Byzantine armies in the eleventh century AD. No other weapon launching heavy missiles has dominated western warfare for so long.

Figure 6 - Schramm's fine reconstruction of the polybolos (photo: Dietwulf Baatz)
I could never understand why Schramm had not been able to get the chain drive of the *polybolos* to work and had substituted bicycle chains and gears. Looking at his original chain drive in this display in the Saalburg Museum, I suggest that his links were too small and would be liable to slip. See Figure 28 for a close-range impact of one *polybolos* bolt into the rear of another.

Chapter 2

Greek origins

'For the katapeltikon *was invented at this time in Syracuse, as a result of the best craftsmen being assembled in one place from all over the world.'*
(Diodorus Siculus 14.42.1)

'Catapult' is derived from the Latin *catapulta*; but the word, like the machine, is Greek in origin: originally spelt *katapaltes*, a noun derived from *kata* 'down' and the verb *pallein* 'to hurl'; later the spelling *katapeltes* became standard. It is used to describe a missile-projecting machine for knocking down personnel and structures, and one which employs mechanical means to produce a far superior performance to that achievable by hand-operated weapons like the bow or the sling. Diodorus' *katapeltikon* appears to be an alternative for *katapeltes*.

No evidence exists to contradict Diodorus' statement that the catapult was invented in 399 BC by Greek engineers working for the wealthy tyrant Dionysius I of Syracuse, during a period of intense warfare between the Greek cities of Sicily and Italy and the Carthaginians. Dionysius equipped his army with weapons whose superior firepower outranged and demoralised his enemies. By offering high wages Dionysius I had attracted to Syracuse large numbers of expert craftsmen who produced '*...all types of catapults* (katapeltai) *and a large number of other missile weapons*' (Diodorus 14.43.3).

The use of the plural '*catapults of all types*' probably refers to the various versions of the *gastraphetes* ('stomach-bow'), the first of two major improvements on the performance of the hand-bow. Biton 61-67 (Marsden 1971, 75-77) mentions two stomach-bows designed by Zopyrus of Tarentum (Taranto in Italy), who may have been working just before 399 BC, and may even have been a member of Dionysius' team. The name 'stomach bow' records the fact that the weapon was initially loaded by using stomach pressure (Figure 8). The name survived even when the bows were greatly enlarged and required a winch to load them. An advanced winched version by Zopyrus (Biton 62-4) was mounted on two support stands and had a bow 2.77m long. It launched simultaneously two bolts 1.85m long and 3.6cm in diameter, indicating that Greek engineers were under instructions to increase the volume of missiles as well as their size, range and weight. These stomach-bows were capable of launching much heavier bolts than the arrows of the legendary Scythian bow and probably outranged it with bolts of similar weight to the Scythian arrows; the stone-throwing versions easily surpassed the impact of sling shot, though probably not the range. With the invention of the winched *gastraphetes* the art of war had entered a new phase.

There is a limit to the pull that can be achieved by an archer's arm, chest and shoulder muscles. Most modern archery enthusiasts use a bow with considerably less than a 45kg pull. The skeletal remains of the *'massively boned'* longbow archers on Henry VIII's flagship, the Mary Rose, have been analysed as capable of exerting a pull of 45 to 81kg (Hardy 1992, 201). Sir Ralph Payne-Gallwey estimated that a 68 to 72kg pull was required to draw the most powerful Turkish composite bows (Payne-Gallwey 1903, Appendix, 21).

It was the ancestor of these Turkish bows that was in standard use by most armies in Dionysius' day. The Greeks had gained their knowledge of this bow from their Black Sea trading contacts with the Scythians, whose skill as archers became famed and feared throughout the Greek world. It was constructed of an inner layer of horn, a central core of wood, and an outer layer of sinew. The action of drawing the bow compressed the horn and stretched the sinew. The combined properties of the three materials made the bow so powerful that when unstrung it sprang back to almost a reverse profile (Figure 7). The widespread use of this bow throughout the Mediterranean world meant that it was very difficult for any army to gain an advantage over its opponents through missile power, except by producing superior numbers of archers and longer-lasting salvoes of arrows.

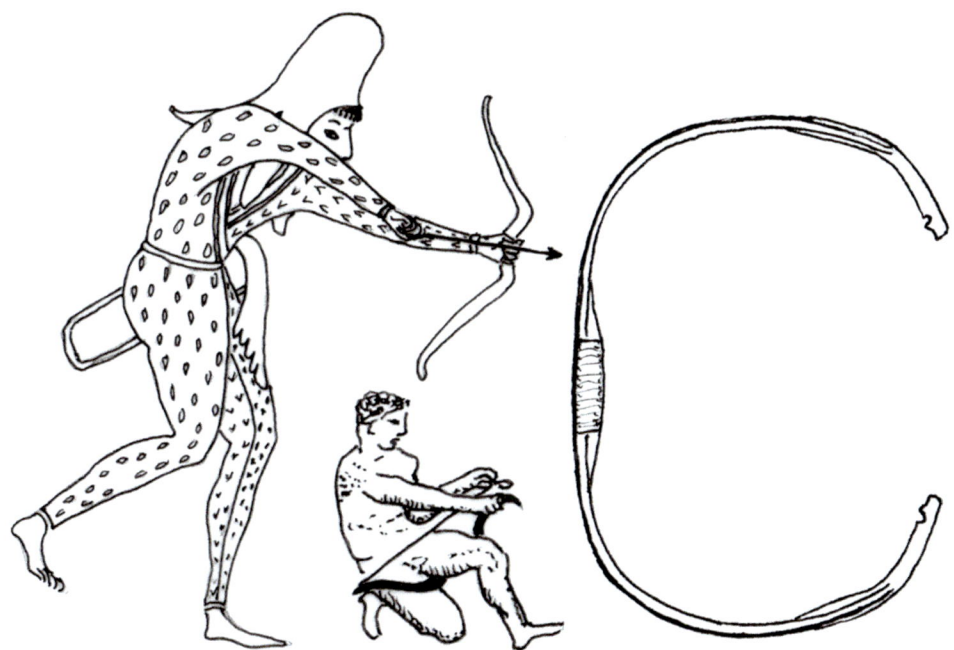

Figure 7 - Scythian archer. Composite bows
Scythian archer (from an Athenian vase painting). Stringing a composite bow (from a coin of Thebes). A Turkish composite bow unstrung in the Doges' Palace arsenal, Venice (from the author's photograph).

The *gastraphetes* ('stomach-bow'): a semi-robotic archer

The *gastraphetes* was a version of the composite bow enlarged to increase its range and strike-power beyond those of a hand-drawn weapon. Schramm's reconstruction (Figure 8 left) uses a bow of spring steel which did not appear until the fourteenth century AD.

Greek engineers devised an ingenious robotic replacement for the archer's arms, hands and fingers (Figures 8, 9 and 10). The left arm that extends the bow was replaced by a long wooden stock consisting of a case and a wooden slider which moved in a dovetailed groove on top of the case. On top of the slider was a missile groove to guide the arrow. The grip of the left hand on the bow was replaced by metal clamps that fastened the bow to the front of the case. Drawing back the bowstring was achieved by means of a metal trigger fitted at the rear of the slider, copying an archer's two fingers on the bowstring. The action of this finger-like trigger in locking onto the bowstring makes it possibly the earliest known cybernetic device in western technology, a device which mimics an action of part of the human body. The operator drew the bow by extending the slider, resting its end on the ground, locking the trigger onto the bowstring and using his bodyweight to push with his stomach against a curved stomach bar fastened to the rear of the case (Figure 8 left). As the

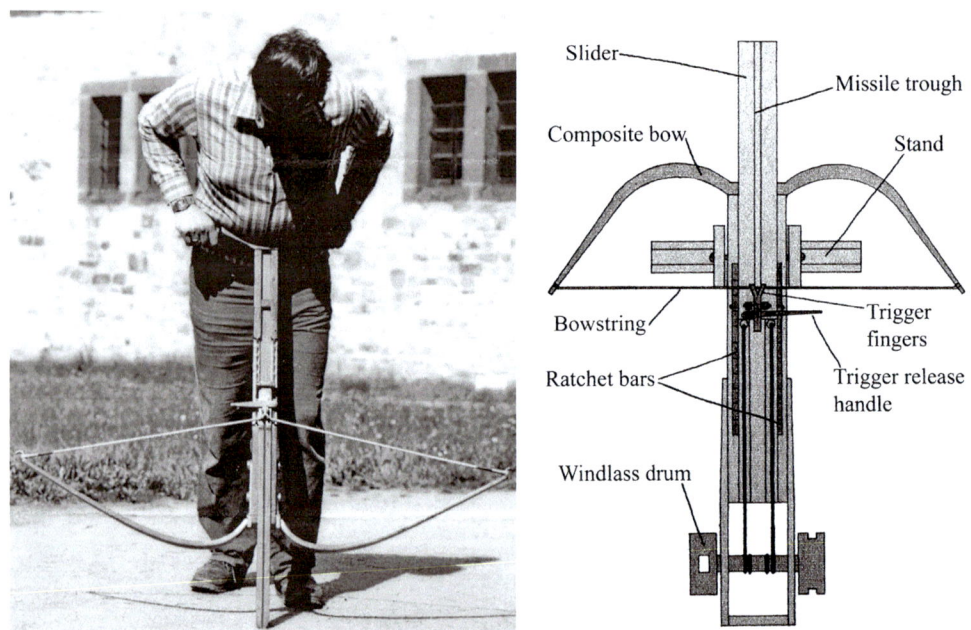

Figure 8 - The stomach bow
(Left) Spanning Schramm's reconstruction of a stomach-bow. Note the operator's grip on the end handles of the curved stomach-bar, and the help afforded by his overhanging stomach (after Schramm 1918/1980). **(Right)** Enlarged, winched composite bow from above (after Marsden).

GREEK ORIGINS

Figure 9 - Testing a gastraphetes
About to release the trigger of an experimental *gastraphetes* at the 2016 Feeley Memorial Shoot. The hand grip on the end of the standard curved stomach-bow loading-handle is clearly visible. Compare this with the handle drawn in red at the bottom of the manuscript diagram in Figure 10.

case was forced downwards, pawls on either side of the slider moved backwards along two toothed ratchet-bars on the case, locking into the ratchet slots as the weapon was spanned. An average person might achieve 68kg stomach pressure, someone heavier might be expected to span a weapon of 84 - 91kg.

Figure 10 is the Paris Codex P drawing of a *gastraphetes* in Heron's *Belopoiika*. The main features of bow, bowstring, case and stomach-bow loading-handle are easy to spot, but the rest require careful matching of the labels of the parts to Heron's text (in Marsden 1971, 21-23). Even then, centuries of copying by scribes who had never seen a stomach-bow have resulted in errors and distortions. Drawing to scale does not appear to have existed. Also, there was no drawing technique for showing an object from two different angles: if needed, the detail of the object's side would be pulled round by 90° and drawn alongside the view from the front. These problems have to be faced when attempting to understand Heron's diagrams of the stone-thrower (Figure 102) and those by the author of the *Cheiroballistra* arch strut bolt-shooter (Figure 77). Unfortunately, there are blank pages where the diagrams should be in the manuscripts of Vitruvius' descriptions of bolt-shooter and stone-thrower.

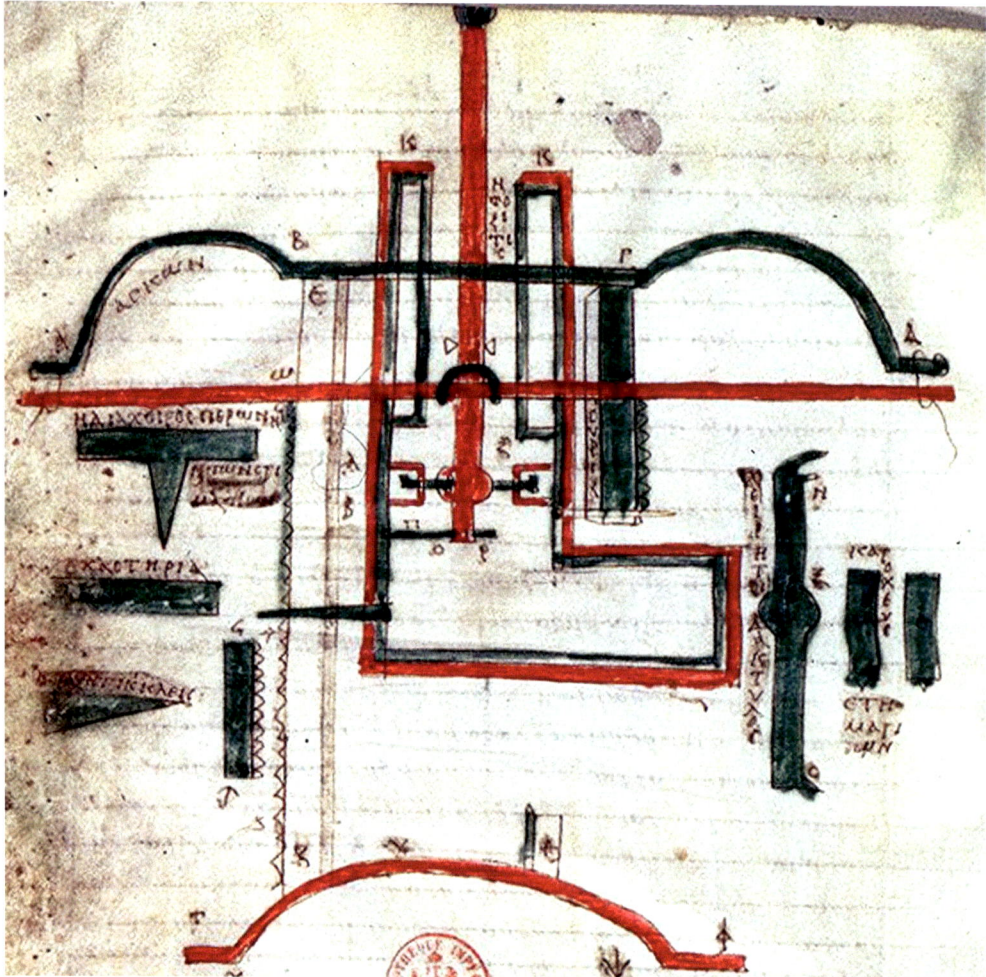

Figure 10 - Codex P drawing of a gastraphetes (photograph: Wikipedia commons)

The introduction of the winch and stand

Bruised stomachs apart, there was clearly a limit to the power of the pull-back achieved by leg power and body weight; therefore, the curved stomach bar was at some point replaced by a winch, enabling the artilleryman to exert almost unlimited pull-back on the bowstring, thus allowing bows of remarkable size to be produced, capable of hurling stones as well as heavy arrows. Heavy duty stands were devised to support the catapults at a convenient height for loading and aiming.

The introduction of the winch also solved a problem which has affected archers throughout history: that the power of the bow is limited by the physical strength of its operator, and its accuracy on the skill achieved by expert coaching and long practice. The late Roman writer Vegetius (Book I, 15) states that, *'About a third or quarter of recruits, those who prove to have more aptitude, should be given regular training using wooden bows and dummy arrows...Instructors should be chosen for this training who are experts... This skill at archery needs to be learnt thoroughly and maintained by daily use and practice'*.

These stomach-bows were capable of launching much heavier bolts than the arrows of the legendary Scythian bow and probably outranged it with bolts of similar weight to the Scythian arrows; the stone-throwing versions easily surpassed the impact of sling shot. With the invention of the winched *gastraphetes* the art of war had entered a new phase. Curiously, the term *gastraphetes* was retained for these catapults even though they were not loaded by stomach pressure. It is easy to visualise the consternation and carnage among Dionysius' enemies, such as the unfortunate Carthaginians at the siege of Motya in 397 BC, who knew nothing of his new bolt-shooting weapons and believed that they were well out of range.

A winch with hand-spikes acting as long levers can be wound back by artillerymen of varying physical strengths and ages (Figure 38). The author's winch-powered reconstruction of the later Roman *cheiroballistra/manuballista* (Figures 60 and 86) has been regularly operated by 'artillerymen' of all sizes and ages from ten to 90, and its draw weight is 335kg. Furthermore, accurate aiming and confident handling of a catapult can be mastered after a comparatively short period of initial training. Catapult winches operated by straight handspikes (capstan bars) are clearly shown on Trajan's Column and on the mss drawings in Heron (Figure 11). An intriguing two-handle winch, apparently operated by a two-handed see-saw action which speeded up loading, is shown on the Cupid Gem (Figure 85) possibly dating to the second century AD.

Even faster loading was achieved by the crank handle, the use of which in Roman times has often been doubted. The earliest use of the crank handle is claimed to be on a flywheel on one of Caligula's ships on Lake Nemi, apparently part of the bucket chain bilge pump (Ucelli 1950, Figure 199 and Inventario 409); but the authenticity of this find has been disputed. Amongst the Roman military items ploughed up in 1962 from a field at Elenovo in Southern Bulgaria (Figure 87) one has been identified, correctly in this writer's opinion, as a combined crank handle and handspike, associated with a catapult field-frame and a possible arm. (Kayumov and Minchev 2010, 336-8 and Figure 14). These items probably date from the 2nd century AD. The fact that this is a combination of a handspike and a crank handle makes its identification as from a catapult certain, because the extra leverage of a handspike is essential to achieving the final few centimetres of pullback. Our tests confirm that the crank handle shortens by one third the time required to winch back the arms and launch the missile.

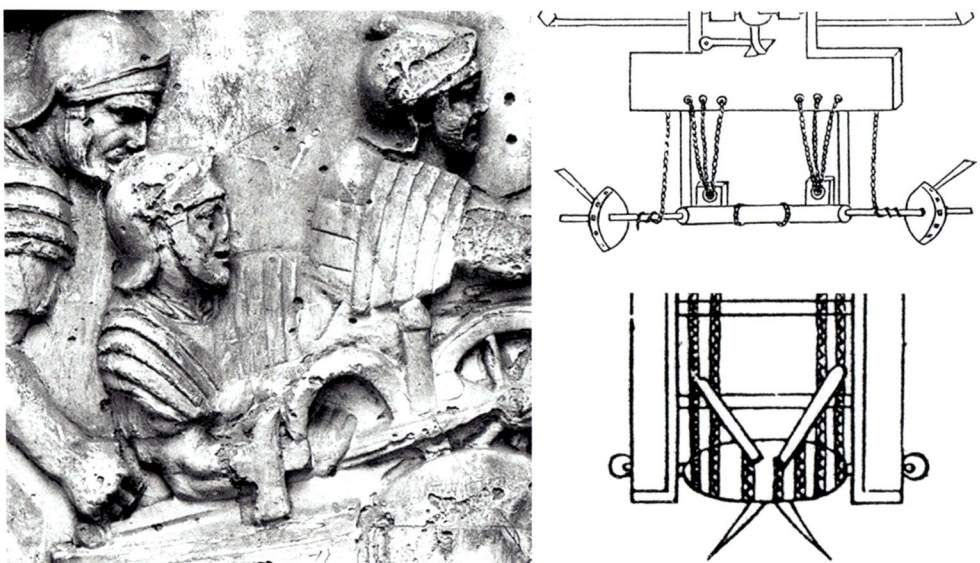

Figure 11 - Catapult winches
(left) Cast of Trajan's Column Scene XL: the rear of the two legionaries manning a cart-mounted arch strut bolt-shooter is holding the handspike of a winch whose round drum is visible. **(top right)** Ms drawing from Heron's Belopoiika of the winch of a stone-thrower. The two outside drums have multiple sockets for the handspikes. **(bottom right)** A different ms drawing from Heron's *Belopoiika* with a single inside drum pierced by fixed handspikes. (diagrams from Carle Wescher's edition).

To return to the Greek development of the stomach-bow: Biton 62-4 describes a winched version designed at Miletus in Asia Minor (western Turkey) by Zopyrus of Tarentum (modern Taranto on the heel of Italy), which was mounted on two support stands and had a bow 2.77m long. Zopyrus' machine launched simultaneously two bolts 1.85m long and 3.6cm in diameter, indicating that the engineers were under instructions to increase the volume of missiles as well as their size, range and weight. Biton 65-7 describes Zopyrus' mountain stomach-bow, with a '*bowl-shaped bracket*' on top of the larger of the two support stands: this sounds like a ball-and-cup joint which would allow rapid swivelling of the catapult to deal with a fast-moving enemy. It is interesting to note that Zopyrus' services were in demand by such widely separated cities as Cumae (Bay of Naples) and Miletus. Greek artillery engineers became highly valued specialists who could command large fees for their expertise.

There is no evidence for the ranges achieved by Dionysius' catapults, but the startling impression they made on those who witnessed their power is evidenced by Plutarch's story of the Spartan king Agesilaus. On seeing a piece of artillery brought over from Sicily for the first time the king exclaimed '*By Heracles! A man's courage in battle is no use any more.*' The occasion was probably in 368/367 BC when Dionysius sent troops to help the Spartans defeat the Arcadians and Argives in a victory described by Xenophon as '*tearless*'. '*Tearless*' suggests a relatively bloodless victory achieved because the new catapults enabled the Spartans and Syracusans to shoot up the enemy while staying out of range.

Torsion-powered artillery

'*Seeking to make the arms of the bow more powerful, but being unable to achieve their objective by means of the horned* [composite] *bows, they made the arms of stronger wood and larger than the arms of a bow; they assembled a frame of four stout beams, wound sinew-rope round the horizontal beams...pushed one of the arms through the middle of the sinew-ropes...*' Heron, Belopoiika 81.

This is the beginning of Heron's description of an early version of a torsion catapult. Two separate wooden arms were inserted into two vertical skeins or springs of sinew-rope mounted in a strong frame of hardwood reinforced with iron plates. Each strand of the rope was pre-stretched by winches around top and bottom iron washer-bars. These washer-bars were slotted into revolving bronze cylinders/washers that allowed the skeins to be twisted with a spanner, forcing the bow arms forward. This twisting or torsion of the rope springs was further increased when the arms were drawn back by winch, storing a massive amount of energy in the sinew-rope.

There is no evidence for the date of its first appearance or which engineer(s) invented it, but this complex and powerful design, using components under great stresses and employing a novel source of energy, is unlikely to have sprung suddenly from some workshop in fully developed form. Lengthy, expensive experiments and trials would be required, giving support to Eric Marsden's belief that circumstantial evidence points to its initial development by engineers working for Philip II of Macedon: he, like Dionysius, possessed the required resources of money, manpower and access to expertise. It is likely that the various versions of the *gastraphetes* continued in use alongside torsion catapults for a long time.

Rapid improvements were made to the torsion design by the engineers of Alexandria and Rhodes, so that by 300 BC catapults could project lethal bolts from two spans (46cm) and three spans (69cm) in length, up to a hefty man-size spear four cubits (1.84m) long; and stone shot could be hurled weighing up to a staggering three talents (78.6kg). Doubts have been expressed as to whether stones as heavy as 78.6kg could have been shot from torsion catapults; but stone balls of this calibre have been found at Hellenistic sites such as Pergamum (Turkey) and Rhodes. Demetrius Poliorcetes in 307 BC fitted the three talent size into the lower floors of his 41m high siege tower at Salamis on Cyprus (Marsden 1969, 105). Philon of Byzantium (51, 21-6) lists eight sizes of stone ammunition from ten *minai* (4.3 kg) to three talents. Vitruvius, Augustus' catapult engineer, has a more detailed list (given above in *Weights and Measures*) which starts at two *librae* (two Roman pounds, 0.65kg) and goes one beyond 240 *librae* (three talents) to a 360 *librae* (118kg) stone. The largest shot in regular use by the Roman army seems to be the one talent (26.2kg), found on sites from Israel to Britain. A one talent machine was recreated by the BBC (Figure 13. See Chapter 8 and Figure 111).

By the middle of the third century BC catapult design had become so thoroughly refined and proven that Alexandrian engineers had been able to devise a set of formulae for calculating the ideal size of every component, based on the length of the bolt or the weight of the stone shot. For the bolt-shooter, the diameter of the rope-spring/washer aperture (Latin *foramen*) was one ninth the length of the bolt, and the measurements of all components were given in whole *foramina* or fractions of a *foramen*. The highly sophisticated formula for the stone-thrower was $D = 1.1 \sqrt[3]{(100 M)}$ where D = the diameter of the rope-spring/washer aperture in dactyls and M = the weight of the proposed stone shot in Attic *minai*. D or fractions of D provided the measurements for all the stone-thrower's components (see comments in Chapter 8).

Figure 12 - Marsden's three-span torsion bolt-shooter (Photograph: Eric Marsden)
Operator's view of Eric Marsden's torsion catapult, shooting three hand-span (69cm) bolts. Eric built this machine in less than three weeks for the BBC programme on Maiden Castle (Figure 21). Hence there was not enough time to have authentic washers cast or to clad the frame with plating.

Greek origins

Figure 13 - The giant BBC ballista, the one talent stone-thrower created for the BBC programme 'Building the Impossible'.

These stone missiles could be projected to ricochet to ranges greater than pebbles or lead bullets propelled from a sling. Unlike catapult bolts which had a single impact, the stone balls were capable of bouncing beyond their first landing and fragmenting into shrapnel if the ground surface was hard enough. The force of their impact could kill or cripple personnel or animals even towards the end of their travel.

The BBC *ballista* weighed ten tons, and was lifted into position on its stand by the Roman system of A-frames, with ropes pulled by teams operating two large horizontal windlasses. Our attempt at building this one talent stone-thrower left us in awe of the massive and complex engineering which coped with compressional forces measured, according to the BBC engineer, in hundreds of tons, and it confirmed our respect for the Roman legions' mastery of the supply and construction of huge volumes of timber, metal and sinew-cord. *Ballistarii* ('artillerymen') were among the specialist soldiers in a legion who were excused fatigues such as guard duty. The erection of such a monster machine must have spread terror amongst the defenders of a city, as did Edward the First's giant 'War Wolf' trebuchet, which so terrified the defenders of Stirling that they offered to surrender without a fight.

Chapter 3

The menace of the new weapon

If the superiority of stomach-bows over conventional hand-bows was remarkable, the increase in range and power achieved by the fully-developed torsion catapults was sensational. There was a rush by non-Greek countries to acquire them, mirroring the present-day arms race to acquire more advanced rockets and atomic weapons. Catapult engineers were in universal demand.

Roman armies encountered them in the early third century BC through contact with the Greek cities of Southern Italy and Sicily. To gain a decisive advantage over the maritime power of Carthage in the First Punic War (264 to 241 BC) they were compelled to acquire two unfamiliar weapons and the skills to use them, warships and artillery, both with the help of Greek engineers. There is one piece of important evidence for the numbers of catapults available to one army. Livy records that during the Second Punic War (218 to 201 BC) Scipio captured the main Carthaginian base in Spain at New Carthage, modern Cartagena. He discovered the following artillery in Hannibal's <u>reserve</u> arsenal:

'120 *catapultae* of the largest kind
281 *catapultae* of the smaller kind
23 larger *ballistae*
52 smaller *ballistae*
    A vast number of larger and smaller *scorpiones*, armour and weapons'
(Livy, *History of Rome* XXVI, 47, 5-6)

Catapultae are bolt-shooters, *scorpiones* are the smallest sizes of bolt-shooters, and *ballistae* are stone-throwers. (See discussion of scorpions in Chapter 5).

Without counting the scorpions, bolt-shooters outnumbered stone-throwers by more than five to one. A similar ratio is found in the work by the late Roman writer Vegetius, who says that there were 55 catapults and ten *onagri* (one-arm stone-throwers) in each Roman legion. The reason for the disparity is that stone-throwers were constructed to project far heavier missiles and were therefore heavier, more complicated and expensive machines, requiring more spring-rope and larger teams of operators. Only the smallest calibres of stone-thrower were light enough to be manoeuvred on the battlefield.

The 120 catapults of the largest kind were probably shooting bolts three or four cubits long (138cm or 185cm), the 281 of the smaller kind two-cubit bolts (92cm). It is almost certain that larger scorpions were three-spans, shooting bolts three hand-spans long

Figure 14 - Carthage stone shot
Large, well-finished stone shot from the Archaeological Park, Carthage (Tunisia). From their sizes (note the size 8 sandal) they are estimated as weighing 1½ talents (39 kg, the two left-hand ones), 1 talent (26 kg, the right-hand pair), and ½ talent (13 kg, the smallest stone). They may date from the destruction of Carthage by the Romans in 146 BC.

(69cm), and that the smaller scorpions were two- span / one-cubit catapults whose bolts were 46cm long.

In 149BC, the city of Carthage surrendered its artillery to the Romans, totalling 'up to 2,000 bolt-shooting and stone-throwing catapults' (Appian, *Lybike* 80), another indication of the very large numbers of catapults involved in the warfare of that time. 5,600 catapult balls have survived to this day on the city's site. Perhaps the Romans had judged that the ammunition was no threat once the catapults had been removed; possibly the weight of such a large number militated against easy transportation. 3,500 of them were found to weigh between 5.0 and 7.5kg. Vitruvius' list would place these as suitable for 15 *librae* to 25 *librae ballistae*. 900 lighter stones weighing from 2.5 to 4.5kg might be intended for 8 to 12 *librae* machines. 650 'heaviest' stones are recorded as between 16 and 40.5kg; 16kg is slightly heavier than Vitruvius' 40 *librae* half-talent machine, 40.5kg matches Vitruvius' 120 *librae* one-and-a-half talent *ballista*. This tallies with my estimates for the weight of shot in Figure 14 above based on comparison to the 26.2kg one talent limestone shot made by sculptress Carol Kirsopp for the BBC (Figure 113) and now a decorative feature in our garden.

Some of these Carthaginian catapults may have still been in service in Marius' day when he is recorded (Sallust *Jugurtha* 94.3) using catapults, archers and slingers to give covering fire for a tortoise formation assault. '*He personally went outside the siege-sheds, set up the tortoise formation, moved to the attack and at the same time terrified the enemy with long range shooting with catapults, archers and slingers*' ('*...et simul hostem tormentis sagittariisque et funditoribus eminus terrere*'). Clearly Republican generals understood the importance of combining catapults with all other methods of launching missiles. This tactic became the cornerstone of Roman firepower, as later emphasised by Hadrian's general Arrian's orders for the battle against the Alani (Chapter 11): '*...the catapults are to fire both bolts and stones, the archers arrows, the spearmen are to launch their missiles... Let stones rain down on the enemy from the allied troops on the higher ground... It is expected that the indescribable volume of missiles will stop the charging Scythians from coming too close to our infantry line.*' A full artillery barrage would include all these other missiles: stones launched from throwing sticks or by hand, javelins and spears, slingshot, and arrows from the bows of foot or mounted archers.

There is no evidence that Rome held a permanent arsenal of catapults that Republican generals could draw on. Marius, Sulla, Pompey, Caesar, and Octavian followed the example of Scipio Africanus and requisitioned or forcefully acquired catapults from neighbouring Greek cities, and set up field workshops to construct artillery with help from available specialist engineers. *tormenta machinasque et advexerat secum, et ex Sicilia missa cum commeatu erant; et nova in armamentario multis talium operum artificibus de industria inclusis fiebant* (Livy xxix 35. 8). '*[Scipio] had not only brought with him catapults and machines but also others had been sent from Sicily with the main supplies and new ones were being constructed in a workshop by numerous craftsmen specialists in such work who had been purposely brought in.*' As Marsden suggests (1969, 176), some of the artillery Scipio had brought with him may have originated from the capture of Hannibal's reserve arsenal in New Carthage.

Sulla's artillery in sieges: the Pompeii 'target'

Archaeological evidence gives the size of stone-throwers used by the Roman general Sulla in his sieges of Pompeii and Athens. Mike Burns has investigated the evidence from Sulla's siege of Pompeii in 89 BC (Burns 2004). By matching the *ballista* balls found at the site with the deep indentations in the city wall (Figure 15 left below), Burns has concluded that Sulla was using stones weighing 15 Greek *minai* (6.54kg = 20 Roman *librae*), while the Pompeians were defending the city with larger machines of 20 *minai* (8,72kg = 30 Roman *librae*) and 30 *minai* (13.08kg = 40 *librae* = a half-talent). Clearly Sulla was using more portable machines, and the Pompeians heavier ones suited to static defence. In addition, there are the impact marks on the walls of lead sling bullets, of which four sizes were found, marks that '*seem to imply that the Romans, and probably the Pompeians, were firing rapidly and with little accuracy*' (Burns 2004, 5). Volume of missiles was the priority.

As to the possible impact of bolt-shooters, Burns reports,'*Some deep and narrow impacts were located which penetrate the wall up to 15cm and might have been the work of bolt shooters. What is interesting about these penetrations is that it is possible to trace the angle at which the missile made contact with the wall, and thus determine the trajectory the Romans were firing from*'. He must be right: only catapult bolts with their iron heads could have penetrated to this remarkable depth. In Burns' Figure 14b and my Figure 16 (left) many small nicks on the surface of the wall are visible. Could these be strikes from arrows?

It seems unlikely that there will ever be a more authentic proof of the impact of Roman stone-throwing and bolt-shooting catapults and of sling bullets than these Pompeian walls. First, it does not take much imagination to recreate the terrifying sound of the travel and impact of the hailstorm of missiles. Second, if these are the depressions in admittedly soft stone, what would the impact and penetration have been on human flesh? Burns points out that the inability to see the approach of slingers' smaller missiles meant that '*The psychological fear of being hit could be just as effective a weapon.*'

The report is intrigued by '*a number of American-football-shaped impacts which measure 14cm long by 10cm wide and have penetrated the wall to a depth of up to 8cm*'. He suggests that because the outline of these impacts is shaped like the lead bullets, they may be caused by Sulla using large lead shot fired by artillery. However, his photograph shows the impact outline as boat-shaped, i.e. with a round 'stern' shaped left end but a pointed 'bow' shaped right end. To me this looks like a glancing impact from a *ballista* ball which has struck the wall at an angle from the left rather than straight on, and has skidded off to the right out of a progressively shallower crater.

Figure 15 - Sulla's artillery: the Pompeii 'target' (Mike Burns. From JRMES 14/15, 1-10)
(left) 15 mina ballista ball matched to impact hole. **(right)** Medium calibre sling bullet matched to impact hole.

Figure 16 - Sulla's artillery
(left) The Pompeii 'target'. Damage to the perimeter wall of Pompeii between the Porta Vesuvio and Tower X by Sulla's army during the siege of April 89 BC (photograph: the author). **(top right)** Excavations at the Piraeus, Athens, have uncovered some of Sulla's *ballista* balls from his siege of 87 BC. Archaeologist George Steinhauer is standing next to a group of them. **(bottom right)** A Sullan artilleryman has vandalised a memorial column to a Greek with a name ending in -lochos, perhaps Archilochos (photographs by kind permission of Wild Dream Films Ltd).

At the siege of Athens in 87 BC Sulla was using much larger *ballistae*, as the stones from the Piraeus Cemetery in Figure 16 right shows. In both sieges the ammunition was cut from local stone, dark grey lava stone and limestone at Pompeii and limestone at Athens.

These Sullan missile strikes have dropped short of their intended target of Pompeii's inhabitants and houses. Excavations behind this stretch of wall have located many damaged properties, large numbers of stone balls and lead sling bullets. The small oval marks are from slingshot, the very small nicks from catapult bolts (see Figure 16 left). In Athens Sulla's troops desecrated limestone columns from the nearby Athenian cemetery by turning them into *ballista* ammunition (Figure 16 right). The largest appear to be 60 *minai* = 80 *librae* = one talent (26.2kg) in weight. Sulla had arrived at Athens without any artillery. He had to borrow some machines from Thebes, and set up workshops at nearby Eleusis and Megara (see Marsden 1969, 110-112). The inscribed column turned into a missile retains the letters *lochos* which is the Greek word for the Latin *centuria*, the century of a legion (see the centurial labelling on the inscribed balls from Egypt in Chapter 10). The desecration of memorials to the dead indicates the ruthlessness of this siege, also the fact that Sulla's borrowed catapults arrived with an inadequate supply of missiles. The problem of transporting the sheer bulk and weight of stone ammunition, even for one *ballista*, must have made it inevitable practice to carve most of the supply from local stone. Compare the vivid picture in Chapter 10 of legionaries toiling in the Egyptian climate at Qasr Ibrim to cut ammunition from the local unsuitable friable sandstone. Sulla's other impact on the town was to recreate it as *Colonia Cornelia Veneria Pompeianorum*. Ambitious generals like Sulla continued to raise and disband armies as required; so, there were no permanently available stores of artillery machines. It would be another half century before Augustus created a professional standing army and appointed *arcitectus* Vitruvius and three others to supervise the construction and repair of catapults for the legions.

Caesar's artillery

One morning in late August 55 BC warriors lining the white cliffs of Britain's southern shore watched the approach of Caesar's large invasion fleet of several warships and about 80 transport ships carrying a force of two legions. The warriors' command of the shore persuaded Caesar to move seven miles to the east to find a more favourable spot for a landing on *'an open and level beach'*. Caesar makes excuses for his troops' reluctance to jump down in armour into the heavy channel swell to face the Britons, who were boldly hurling their missiles from the beach and from horseback as they rode into the water. His solution was to order his warships to row quickly out of the mass of transports to a position on the exposed flank of the enemy and drive them away *'with slings, arrows and torsion machines'* (*fundis, sagittis, tormentis*). This manoeuvre *'was of great help to our men,'* says Caesar, *'for the natives were terrified by the shape of the ships, the motion of the oars and the unfamiliar type of torsion machines; they halted and retreated, but only a short distance'* (*Gallic War* IV, 25). It required the courage of the eagle-bearer of the Tenth Legion to shame his colleagues into jumping out of the ships and engaging in the fiercely fought battle for the beach.

The artillery (Caesar uses the word *tormenta* 'torsion machines') may have been projecting 69 and 92cm bolts from three-span and two-cubit catapults and stones of about grapefruit size, hitting the Britons at ranges well beyond the accompanying showers of slingshot and arrows. The catapults would have gained additional range because they were almost certainly mounted high up on the warships' towers.

Catapults were regularly used by Roman armies in combination with slingshot, arrows, spears and stones thrown by hand or with the *fustibalus* arm extension to create a dense, long-range hailstorm causing injury or death over a wide area. The sling bullets striking the Britons on the beach were probably capable of killing at ranges up to at least 80m, and causing serious injuries at twice that range. Modern slingers on Ibiza can hit a one metre square target at 200m (Jobey 1977-8, 87, note 32). See now Keppie 2023, 75-6.

A Roman Scythian/Parthian type of bow is very accurate, according to Vegetius (II, 23), at a target 600 Roman feet (177m) away (see Range page 33). Eastern enemies of Rome continued to use this type of composite bow with great effect, most notably the Parthians, whose unceasing showers of arrows, delivered to the battlefield by camel trains, destroyed the army of Caesar's colleague Crassus at Carrhae in Mesopotamia, modern Iraq, two years after the landing in Britain.

Three years after the British invasion Caesar again depended heavily on artillery to save his army from an adverse situation. At Alésia, Mont Auxois in central France, he confronted the coalition of Gallic tribes led by Vercingetorix. Caesar employed rapidly constructed earthworks to compensate for the fact that he was heavily outnumbered by the 80,000 Gauls under Vercingetorix whom he had surrounded in their hill fort, and the 285,000 Gauls who arrived to rescue them. He describes (*Gallic War* VII, 69 and 73) the construction of eleven Roman miles of inward-facing and fifteen miles of outward-facing palisaded ramparts, with towers every 24m, two ditches and a 'minefield' of sharpened branches, stakes in concealed pits, and iron hooks. Figure 17 shows the wide 'killing zone' of obstacles in front of the palisaded rampart and two ditches, The sharp stakes set in the honeycomb of pits nicknamed 'lilies' are similar to the pits outside the forts at Rough Castle on the Antonine Wall, Scotland and at Wallsend and other sites on Hadrian's Wall. Caesar mentions the large numbers of missiles discharged by his artillery, and describes Gauls shot from the rampart or impaled on the stakes.

Figure 17 - Alésia
Reconstruction of Caesar's siege lines at Alésia (52 BC) in the former Archéodrome archaeological park at Beaune (Burgundy, France). The park was closed in 2005. Fortunately there is now a similar reconstruction of Caesar's siege lines at the on-site museum, the Alésia MuséoParc, at Alise-Sainte-Reine in northern Burgundy (photograph: the author).

The Menace of the New Weapon

The 1991 excavations (Brouquier-Reddé 1999, 277-288) added to the tally of weapons found on the site by Napoleon III's explorations. Two stone catapult balls found near Camp A weigh 4.6 and 7.5kg, and probably come from small to medium-sized stone-throwers of 15 and 25 *librae*. Considerable numbers of Roman twin-barbed arrowheads have been excavated. A group of 24 lead sling bullets includes two that are inscribed with the name of Titus Labienus, one of Caesar's generals, perhaps the commander of that sector of the battle. There is now an excellent on-site museum, the Alésia MuséoParc, at Alise-Sainte-Reine in northern Burgundy. Unfortunately, the museum's reconstructed *scorpio* has undersized washers and lacks plating.

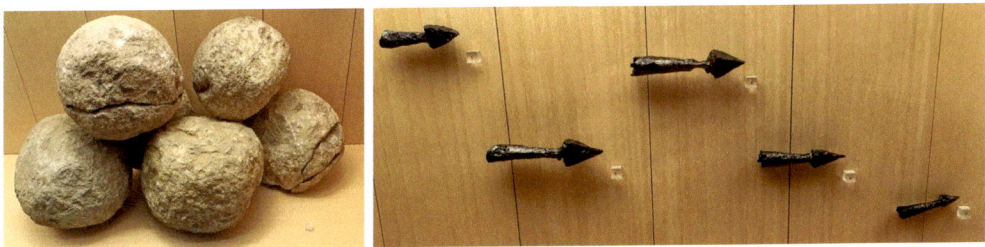

Figure 18 - Missiles from Caesar's siege found at Alésia (Mont Auxois in central France)
These finds of Caesar's Alésia catapult missiles in the Musée d'archéologie Nationale at St Germain-en-Laye were found in Napoleon III's original excavations conducted by Colonel Stoffel in the 1860s (photograph: the author).

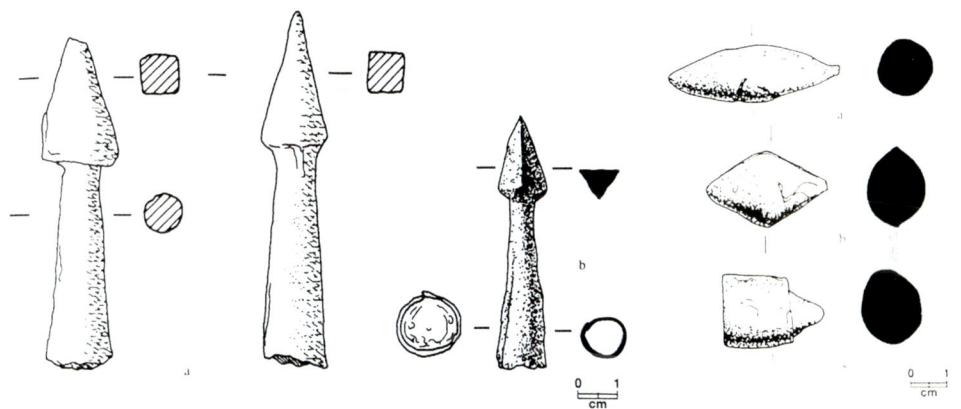

Figure 19 – Boltheads and sling bullets
(left) Two artillery boltheads from the more recent Alesia excavations, with square bodkin points and of a socket size suited to three-span scorpions. **(centre)** A triangular point bolthead, possibly from a two-span scorpion. **(right)** Three sling bullets: Top: lead, olive shaped. Centre: lead, biconical. Bottom: lead, acorn shape. (from Brouquier-Reddé 1997, 280-1)

Chapter 4

The bolt-shooter: accuracy, range and effects

Dramatic archaeological evidence comes from the Claudian invasion of Britain which began in AD 43, during which the future emperor Vespasian, commanding the Second Legion, fought thirty battles, defeated two tribes and captured more than twenty hillforts and the Isle of Wight. Archaeologists believe that they have located evidence of his artillery in action at hill forts in Dorset.

Maiden Castle, Dorset

At Maiden Castle near Dorchester Sir Mortimer Wheeler interpreted his discovery of forty skeletons of Britons buried in shallow graves close to the east gate as a 'war cemetery' and evidence of a great massacre of the British defenders, with the legionaries storming the gate under the cover of artillery fire (Wheeler 1943, 61-3). However, re-examination of the fifty-two skeletons discovered to date reveals that only fourteen have battle wounds and some may not have died at the hillfort (Sharples 1991, 124-5). There is one victim of Roman artillery with the iron point of a bolt in his spine (Figure 20). Wheeler 1943, Plate LVIIIA and page 352, Skeleton P7A 'adult

Figure 20 - Artillery victim at Maiden Castle
Iron Manning Type IIB bolthead from a Roman catapult lodged in the spine of a Briton buried at Maiden Castle, Dorset. Skeleton now in the Dorset County Museum, Dorchester.
(photograph: © The Society of Antiquaries of London.)

male, aged 20-30 years, height 5 ft 5¼ in.'. The bolt appears to have travelled uphill and entered around his belly-button. The Manning Type IIB Roman leaf-shaped iron bolthead has lodged in the Briton's twelfth thoracic vertebra. The 10-11mm internal diameter of the socket is too small for a three-span bolt-shooter and was probably shot from a Xanten-Wardt size of catapult (Figure 23). The staff of Dorchester Museum kindly checked the socket size for me. This head has too small a socket to be a British spearhead as argued by Russell (2019, 329 and Figure 4). In attempting to deny the existence of Wheeler's siege, Russell fails to take account of the scatter of Roman catapult boltheads recorded by Wheeler (text below) in the approaches to the eastern entrance and in three excavated areas on the top. The large numbers of leaf-shaped socketed boltheads found on the Hod Hill site since the nineteenth century and now in the British Museum have been recorded and analysed in Manning 1985, 175-6 and Plate 85. A large percentage of them have this small diameter of socket, as have the heads recorded from Vindolanda by Robin Birley (Birley R. 1996), and those from many other sites: a bolthead of this size has been found in the most recent excavations at Ham Hill, Dorset (John Smith and Marcus Brittain pers. comm.). Compare this leaf-shaped blade in the victim with a similar one from Hod Hill in Figure 27 right and the comments in the caption there.

The square hole in the skull of another victim, Skeleton P7 (Wheeler 1943, 352 and Plate LIII) has now been technically examined (Feeley, Wilkins and Wilkinson: Ancient Discoveries' *Ballistics* TV programme) and shown to have been most likely caused by a *pilum* thrust; a catapult bolt would have shattered the bone around the impact hole. Wheeler's picture of a post-battle cemetery has been disproved, but his report (Wheeler 1943, 62 and 281, Figure 93) records a few catapult boltheads in the east entrance and its outworks and in three excavated areas on the top. His picture of the Roman technique of assault under cover of artillery fire is authentic and a Vespasian speciality. There is an eyewitness description (Josephus, *Jewish War* III, 166-8) of Vespasian clearing enemy battlements with such a barrage from his three legions at the siege of Jotapata (Palestine) in AD 69: '*Vespasian ordered his artillery, numbering a total of 160 machines …to shoot at the defenders on the wall. In a coordinated barrage the catapults sent long bolts whistling through the air, the stone-throwers shot stones weighing one talent, fire was launched and a mass of arrows. This made it impossible for the Jews to man the wall or even the area behind it that was strafed by the missiles. For a mass of Arabian archers, spearmen and slingers was in action along with the artillery.*'

Like most of the enemies of Rome, the Maiden Castle Briton struck by the catapult bolt would not have been wearing body armour. A writing tablet from Vindolanda (Birley A. 2002, 96), critical of British fighting tactics, describes the Britons as wearing no armour (*nudi*). Our tests with a reconstructed bolt-shooter of this period against mail (Figure 52) or plate (Figure 51) show that even if he had been wearing body armour he would have been put out of action.

Figure 21 - Marsden's three-span at Maiden Castle: a television first
Eric Marsden shooting at the ramparts of Maiden Castle, Dorset with his three-span *scorpio*, for the pioneering BBC television programme on the site by Sir Mortimer Wheeler. See also Figure 12. (from Paul Johnstone, Buried Treasure, Phoenix House, 1957, Plate 61).

Hod Hill, Dorset. A bolt-shooters' 'target'

25 km north-north-east of Maiden Castle, at the Durotriges' hill-fort of Hod Hill, unique proof of the accuracy of the Roman bolt-shooter was found when Sir Ian Richmond examined two round huts and a very large palisaded enclosure which had come under attack. 17 iron boltheads were found, most of them still embedded in the chalk surface in the position where they had landed, their wooden shafts having rotted away. Sir Ian conjectured (Richmond 1968, 31-3) that they had been fired in an arc of about 10 degrees, starting with four bolts in the rectangular enclosure outside Hut 36a, and one inside that hut. No less than eleven had landed inside Hut 37, and only one had overshot to land in the adjacent Hut 43; none had undershot into the adjacent area of Hut 56. The slope of the site in the direction from which the bolts came, and the existence of other huts on that alignment interfering with the line of sight, would have made it necessary to shoot from a high viewpoint; Sir Ian suggested a portable siege-tower not less than 50 feet high to overlook the rampart.

The range from such a position would have been about 170m. An alternative scenario that I have proposed below is that the shooting was from a position on the rampart

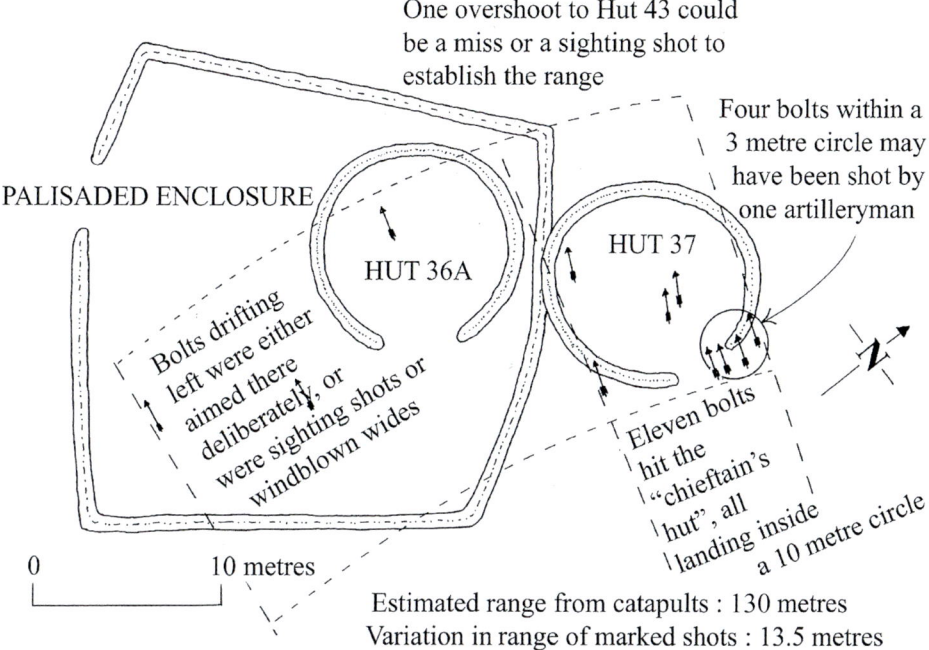

Figure 22 - Plan of boltheads targeting the chieftain's huts and enclosure
The author's analysis of Sir Ian Richmond's Hod Hill 'target'. Sir Ian's plan only marks the missile heads that appeared to be still embedded in their position of impact (after Richmond 1968, Figure.14).

itself, at a range of some 130m. A recent suggestion that these boltheads were shot from the Roman fort in the hillfort's NW corner (Russell 2019, 336) is untenable because the bolts were launched from exactly the opposite direction.

Analysis of the position of these missiles shows accuracy of the kind that will regularly knock out troops and horses at this range, '*demonstrating then and now the deadly and devastating precision with which the ballista could be handled.*' (Richmond 1968, 33). Sir Ian discussed this drawing with me at length, explaining that he had noted the bearings of the boltheads that were still in position embedded in the chalk ground surface, and kept a sharp lookout for other heads in all the huts that he excavated and in the areas around them, the white surface making them easy to spot. The two bolts in the centre of Hut 37 are on a slightly different alignment from the four at the entrance: my belief is that they have caught the left-hand frame of the doorway on the way in and ricocheted.

Sir Ian surmised that the chieftain's hut was selected as an artillery target to intimidate the defenders into surrendering: '*The lack of evidence for an assault on the hill-fort, or for subsequent devastation within it, would strongly suggest that capitulation was then and there induced by showing dramatically what concentrated fire could do.*' (Richmond

1968, 33). There was no sign of victims in shallow graves, as found by Sir Mortimer at Maiden Castle, or of any battle debris in Sir Ian's trenches across the ramparts or at the gates. It is unlikely that this dramatic and attractive scenario will ever be proved or disproved. An alternative possibility which has been suggested is that these bolts were shot at the huts at a later date during target practice by the legionaries living in the Roman fort that was built in the corner of the hill-fort. This seems less likely because of the direction from which the bolts were launched and because after its capture the Romans had levelled the huts in the hill-fort (Richmond 1968, 33): Sir Ian noted a worn coin of Germanicus *'dropped when the timbers of the palisade* [of Hut 36a] *were removed.'* (Richmond 1968, 22).

The palisaded enclosure and Huts 36a and 37

The complex occupies a commanding position on the steep slope that looks down onto the ramparts. Even when the ramparts were at their original height, there would have been a good field of view to the SSE from the doorway of Hut 37 to the high ground south of the present hamlet of Ash (Figure 23 top) and to the south in the direction of the village of Stourpaine. This would have allowed the inhabitants of the huts to observe the fateful arrival and ominous size of Vespasian's army. Conversely Vespasian would have had a clear view of the complex when planning his attack. The high ground to the south would have afforded Vespasian a good initial view of Huts 36 and 37 and the palisaded enclosure, which are at the top of a slope. Information from prisoners and/or activity in and around the enclosure and the huts, perhaps the presence of war-chariots, may have confirmed the importance of this selected target.

There are other huts of similar size visible on the air photograph of Crawford and Keiller (*Wessex from the Air*, 1928, 36 Plate I), but it is the unique size of the palisaded enclosure that was one of the factors that led Sir Ian to decide that it belonged to an *'especially significant personage'*. Dave Stewart's important 2007 magnetometry survey of the whole area inside the Iron Age ramparts confirms the enclosure's uniqueness. Its Hut 36a contained two iron spear heads in the place next to the entrance where other excavated huts, numbers 43, 56 and 146, contained large caches of sling stones: possession of spears is a further indication of high status. O.G.S Crawford pointed out that one of the tracks that start at the main Steepleton Gate heads straight for this enclosure.

Which bolt-shooters were used?

Most of the large numbers of other Hod Hill boltheads drawn and analysed by Professor Bill Manning (Manning 1985, 170-77, Figure 34 and Plates 82-5) have sockets between 10 and 11 mm in internal diameter, the calculated size for the Xanten-Wardt *scorpio minor*, and too small for the 18 - 18.5mm outer diameter, 15 – 17mm inner diameter, attested for the sockets of the three-span *scorpio maior*. The diameter of the bolt for the various sizes of bolt-shooters can be estimated from the size of the exit aperture in their spring frames. This means that the attack was probably mounted not by three-span larger scorpion bolt-shooters, but by one or more smaller scorpions of the lighter Xanten-Wardt size that can easily be carried on one shoulder, even by nonagenarians like myself (with a second artilleryman carrying the collapsible stand and bolts). This allows a slight change to be made to Richmond's scenario, discarding the need for a bolt-shooter to be raised onto a portable siege-tower, the time taken for which would have destroyed the element of surprise.

Vespasian's battle plan

Until the battlefields of Boudicca's defeat and *Mons Graupius* are located, the assault on Hod Hill remains the only identifiable battle site from the invasion of Britain where the battlefield tactics can be estimated with some certainty, whether or not the bombardment of Hut 37 influenced the surrender. The vertiginous drop down into the River Stour on the north side of the hill fort (Figure 26), and the fact that the hill fort only had two gates, narrow down the tactical disposition required to surround and trap the occupants.

Figure 23 - Filming at Hod Hill for Channel 5, June 2012 **(top)** The author and Dick Strawbridge are standing on the hill-fort's rampart at the estimated position of the catapult(s) for the shoot at the chieftain's hut. (photograph: Tom Feeley) **(bottom)** Tom Feeley arming his reconstruction of the Xanten-Wardt scorpion. (photograph: the author)

The following attempt to reconstruct the attack on Hod Hill must necessarily be conjectural.

Vespasian's personal courage and leadership in battle are frequently vouched for, and make it likely that he led the assault on the main hillfort gate: Suetonius (*Vespasian* IV,6) '...*in the storming of a fortress he was struck on the knee by a stone and received several arrows in his shield.*' Josephus (Jewish War III, 151) records as an eyewitness that '*... when Vespasian had positioned his archers, slingers and the whole of his long-range artillery and ordered them to shoot at the Jews, he himself advanced with the infantry up the slope to the spot where the defensive wall was weak...*'

It is important to note that the assault on Hut 37 was mounted from a position well away from the hill-fort's two gates which the defenders would surely have expected to be the main focus of the Roman attack. In order to seal the defenders in and prevent the emergence of their highly skilled, fast-moving charioteers and horsemen, (Richmond 1968, 9 for an estimate of chariot numbers), Vespasian would have had to position his *Legio II Augusta* in front of both gates, possibly under cover of darkness (Figure 26). Dawn would then have revealed to the unarmoured Britons a sea of iron-clad Roman infantry and cavalry, interspersed with artillerymen, archers, and slingers.

The threat of battles at the two gates would have drawn the Britons' attention away from the catapult attack on Hut 37. Vespasian's briefing documents for his campaign against the tribes of Southern Britain would surely have included the information recorded by Strabo (*Geography* IV, 4, 2) that the whole Gallic race was war-mad and quick for battle, coming out for a fight all together without circumspection, so that they could be easily dealt with by anyone who was willing to use a stratagem.

Vespasian's stratagem could well have been a surprise assault on the huts housing the chieftain, his family and entourage. For the 2013 Channel Five TV programme '*Beat the Ancestors – The Roman War Machine*' I suggested that this assault could have been led by the legion's highly experienced *praefectus castrorum* Publius Anicius Maximus, previously the *primus pilus* of the Syrian *Legio XII Fulminata*. Unfortunately this TV programme was trimmed to fit the time slot, and the discussion of the Durotriges, mention of Maximus and many details of the assault were edited out. The inscription from Antioch set up in Maximus' honour (*C.I.L. iii. Suppl.* 6809) records that he was awarded the *corona muralis* (gold mural crown in the shape of a turreted city wall) by the Emperor (Claudius) for his action in this British campaign. This rare decoration was for being the first to cross the enemy battlements and plant the standard on them, by implication an action that contributed to the capture of the fortress or town. Of course, he could have won it at one of the other sieges of the hill-forts of the Durotriges or Dumnonii, but it fits in well with Richmond's hypothesis that the accurate missile shooting may have caused the surrender.

While Vespasian was threatening an attack on the wooden gates and adjacent ramparts, Maximus could have led a fast-moving small detachment of artillery marksmen, using boarding planks to bridge the two defensive ditches (Figure 24 below), which were quite shallow and far less of an obstacle than those of Maiden Castle. On setting up one or more of the smaller Xanten-Wardt size of scorpion bolt-shooters on the hill-fort's rampart, he would have looked straight across into the entrance of Hut 37 at a range of about 130m. One can only speculate where the tribal chieftain, his family and servants were when the lethal barrage thudded into the huts and the palisaded enclosure.

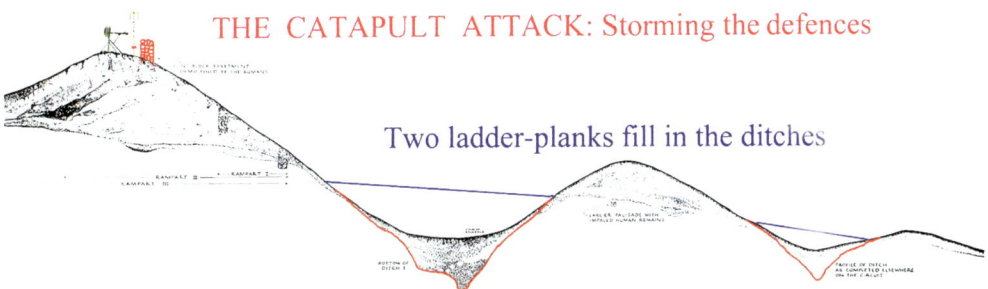

Figure 24 - Vespasian's possible plan of attack – crossing the ditches
The catapult attack group crosses the hill-fort's ditches and sets up the legionary standard and catapult(s) on the rampart. (after Richmond).

Figure 25 - Reconstruction of an Iron Age hut
Reconstruction of an Iron Age hut and adjacent enclosure at the Butser Ancient Farm site. (photograph: the author)

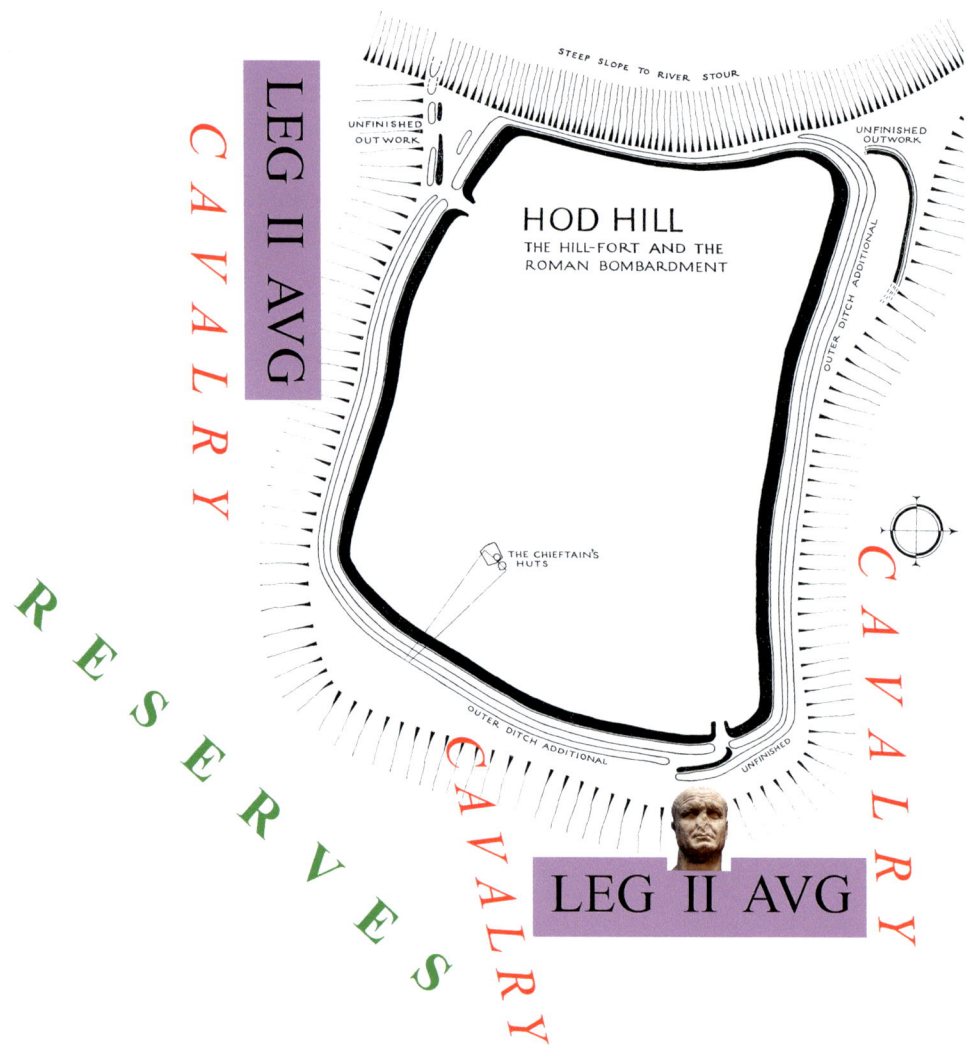

Figure 26 - Vespasian's plan of attack
Suggested positions for Vespasian's assault, blocking the two gates of the hill-fort. Portrait of Vespasian in the British Museum. (diagram after Richmond).

The evidence of the boltheads

The most intriguing feature of Sir Ian's discovery also demands an explanation: why were the eleven bolts left embedded in the chalk in their position of strike? No attempt appears to have been made to recover them by pulling them out by their wooden shafts. It is most unlikely that the shafts had separated from the iron boltheads on impact: in many years of shooting replica bolts we have found that this only happens very rarely, even when bolts penetrate hard targets, such as the replica iron shield bosses backed by 18mm ply pierced in Figure 1. When pulled out from deep penetration of the ground surface, the pinned socketed heads remain firmly attached to the shafts. It is tempting to suppose that the projecting shafts were displayed straightaway to the hill-fort's chieftain and the inhabitants, to drive home the threat of the devastating accuracy and strike power of this unfamiliar weapon. They may have been left on show, partly to continue to trumpet this warning and partly because of the artillerymen's pride in their awesome marksmanship. There is no way of knowing how many other bolts had struck the hut, perhaps lodging in the roof or walls, or hitting its occupants. As noted above, Sir Ian kept a sharp lookout for boltheads in the areas around the enclosure and Huts 43, 37 and 56, but found none.

No finds of *ballista* balls are recorded in the Maiden Castle report, or in the two-volume report on Hod Hill. Vespasian appears to have relied on the bolts of *catapultae*, in particular scorpions of the Xanten-Wardt size, whose velocity, weight, accuracy and extreme ranges easily outclassed the Durotriges' weaponry and made the defeat of all the Iron Age tribes of Britain inevitable. The widespread presence of this smaller scorpion throughout Britain can be traced by the boltheads with this 10-11mm internal socket diameter. For example, many of the boltheads on display in the museums at Vindolanda and Carvoran are of this size.

Of the two Hod Hill boltheads in Figure 27 the nearer one, 8.1cm long, is of the standard type with pyramidal bodkin head. The other has a leaf-shaped blade which would be much quicker to manufacture in an emergency and is the Manning Type IIB bolt which is in the spine of the Maiden Castle victim in Figure 20. Another find from the site in Figure 27 (Richmond 1968, Figure 58, B1a), recorded as a carpenter's bit, found in a pit just outside Barrack 1 of the Roman fort, is a large catapult bolthead of the

Figure 27 - Scorpion and Hod Hill missiles
(left) Len Morgan and Tom Feeley's Xanten-Wardt scorpion, with Vitruvius' curved arms. The 26mm wide aperture in the front plate limits the size of bolt. **(centre)** Two of the socketed Hod Hill boltheads on display in the British Museum's Roman Britain gallery. **(right)** The tanged Hod Hill bolthead referred to in the text. (bolthead photographs by kind permission of the British Museum)

tanged type found at Qasr Ibrim (below, Figure 124) and now being recorded on many sites. This has a length of 118 mm, some 40% longer than the Qasr Ibrim bolthead; one of those from Vindolanda (Birley R. 1996, Figure 13, no. 94) is of a similar size at 116 mm, and weighs 35 gm. Both these are likely to belong to a two-cubit bolt-shooter, which would have had an arm span of about 175 cm and launched a bolt 92 cm long. This should dispel any doubts about whether legionaries were garrisoned in the fort. A third bolthead of this size has been found near Dumfries (pers. comm. Andrew Nicholson) and is now in store at Dumfries Museum.

Artillery platforms (*tribunalia*) at Hod Hill

Sir Ian's excavation of the small Roman fort tucked inside the Iron Age defences examined two platforms next to the East and South Gates, in the position for artillery platforms recommended by the text book on constructing fort defences (*de munitionibus castrorum*, 58, 'you should remember to...construct platforms [tribunalia] *for torsion artillery around the gates...*'). Sir Ian suggested (Richmond 1968, 73) that the two platforms were *ascensus*, sloping access ramps up which the catapults could be manoeuvred onto the ramparts. He describes the south one as '*22 feet [6.7m] in width and 20 feet [6m] from its tail to the back of the 10-foot [3m] rampart, against which it very clearly buts. It is formed by a platform of chalk rubble rising at an angle of 26 degrees from ground level at the back...*' He identified narrow turf-built staircase-ramps on either side of the platform and parallel to the rampart. The 22 feet width of the feature would have allowed more than the pair of catapults suggested by Sir Ian to be mounted on the rampart. If they were two-cubit *catapultae*, whose presence in the fort is vouched for by the large tanged bolthead described above, three could be accommodated with ample space for the crews; and if the other platform at the East Gate had the same allocation this would give a ratio of one two-cubit bolt-shooter for each of the six legionary centuries.

Accuracy

Writers of all periods confirm the accuracy of the bolt-shooters at closer ranges (see below Range: Point 2), which were more suited to precision shooting than stone-throwers. Caesar (*Gallic War* VII, 25) records that at the siege of the Gallic hillfort of Avaricum (52 BC), a *scorpio* killed a Gaul attempting to set fire to the Roman siegeworks, and then scorpion bolts killed a succession of Gauls who picked up the same firebrand. During the siege of Leptis Magna in the Civil War in Africa (46 BC) a decurion commanding a cavalry squadron was struck and pinned to his horse by accurate fire from a *scorpio* (*War in Africa*, 29, 3). The incident not only caused the rest of his squadron to flee back in terror to their camp, but also made them scared of attacking the town again. The fact that these are Roman cavalrymen on the receiving end of a type of Roman missile no doubt familiar to them makes this a particularly impressive example of the panic caused by artillery fire. Livy describes the more

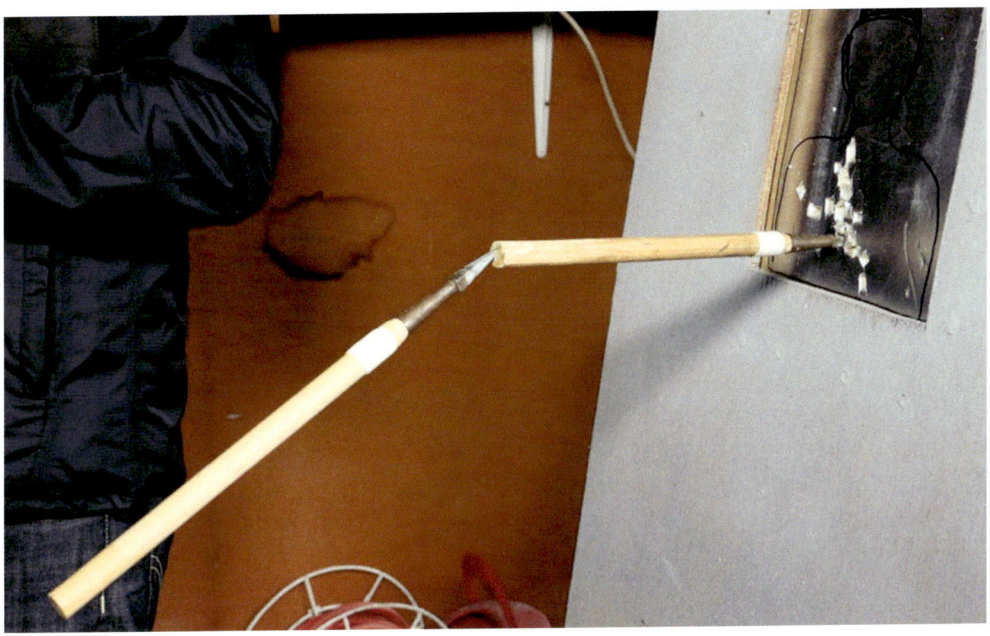

Figure 28 - Close range impact with the Polybolos *(photograph: the author)*
This spread of shots from the Polybolos in Figure 28 occurred during filming for a TV programme in the author's garage. The author was shooting unflighted bolts at a target marked with a human outline, when one bolt stuck in the rear of another. The range was only about three metres, and the catapult rope-springs were only very lightly tensioned for safety reasons. Erwin Schramm had achieved an even more impressive close-range result, splitting one bolt with its successor when demonstrating his version of the *Polybolos* to the Kaiser.

controlled reaction of the Roman general Scipio in 205 BC when attacking the walls of Locri:

> 'Scipio had advanced towards the wall when the man who happened to be standing next to him was struck by a scorpion. The danger implied by this incident alarmed him and he moved his camp well out of the range of missiles'.
>
> (Livy XXIX, 7, 6)

Range

There is a lack of ancient evidence for the ranges achieved by Republican and early Imperial bolt-shooters. Agesistratus' 3½ stades (647m) with a three-span is recorded by Athenaeus Mechanicus (Wescher 1867, 8) with the comment, '...*he so outstripped his predecessors that anyone reporting the facts on his behalf is not easily believed. For his three-span catapult shot three and a half stades with twelve minae of sinew.*' (647m with 5.24kg of sinew rope). Athenaeus implies that to shoot a three-span bolt, 69cm long and probably weighing about 180gm, this far was an exceptional achievement, one which beat normally achieved ranges and was therefore somewhat difficult to believe. Agesistratos was a specialist Greek artillery engineer who no doubt tuned his catapult to perfection and chose the time and weather. He may, of course, have used a flight arrow. This implies that the average legionary artilleryman would achieve a much more modest maximum range, possibly in the order of 400 metres or more. Much would depend on the build quality of each legion's machines, there being no uniformity of performance as in modern factory-produced weapons. Josephus notes (Jewish War V, 269) that the Tenth Legion's catapults at the siege of Jerusalem were superior, '*their scorpions were more powerful and their stone-throwers larger.*' The gradual loss of power from deteriorating sinew-ropes would also affect performance.

What mattered in the end for Roman generals was the ability to outrange the enemies' conventional missiles. Tacitus (*Annals* XV, 9) confirms this in describing Corbulo's success in AD 62 against the Parthians, when he drove them back off the far bank of the Euphrates, crossing the river with tower-mounted catapults on rafts of large boats which had been linked together by beams: '*their stones and bolts travelled further than the Parthians' answering salvoes of arrows*'. The Parthian bows of Scythian type probably had the longest effective range of any weapon fielded by Rome's enemies. Roman units of *sagittarii* used them, and Vegetius (II, 23) says that for practice they and slingers set up bundles of brushwood targets at a range of 600 Roman feet (177 metres), hitting them *saepius* ('more often than not'). When accuracy was less important than range, no doubt the Parthian bow could exceed 200 metres. See the discussion by Ureche 2013.

The metal frame arch strut bolt-shooters introduced at the end of this century increased the arm arc from about 50° to about 70° and would have guaranteed an even greater maximum range. The sixth century writer Procopius (Goth. i. 21. 17 in Marsden 1971, 247) describes their stocky, wooden-flighted bolt being '...LAUNCHED WITH SUCH FORCE THAT IT REACHES NOT LESS THAN TWICE THE RANGE OF A SHOT FROM A BOW ...' The Greek double negative is equivalent to a strong positive: assuming that the '*two bow shots*' range refers to the Parthian/Scythian bow, Procopius implies that the arch strut bolt-shooters achieved at least twice that range, probably around 500 metres or more.

Ranges achieved with modern reconstructions are limited by the fact that they do not use sinew-rope, which the artillery writers mention as the preferred material, superior to horsehair (Chapter 5. 20). Schramm's reconstruction of the three-span Ampurias catapult (Figure 31) armed with horsehair and with an unspecified bolt achieved 305m in 1918, and was still capable of 285m 64 years later. If it had been equipped with Vitruvius' curved arms (Figure 4) allowing greater arm travel and giving the springs more twist, it would have shot further. This gives credibility to the modern estimate of 400m as an absolute minimum range achievable by the wood frame three-span legionary bolt-shooters with springs of sinew.

The Bolt-Shooter: Accuracy, Range and Effects

Tom Feeley's reconstruction of the Vitruvian three-span in the hands of the *Legio VIII* reenactors has achieved ranges up to 365m with the correct length of bolt, using prestretched polyester. Their best rate of launch to date is five bolts per minute. The large four-cubit bolt-shooter constructed by Leo Todeschini's engineering team for the 2012 Channel Five TV programme '*Beat the Ancestors*' (under construction supervision by Tom Feeley and design advice from myself) achieved 318 metres with authentic Vitruvian curved arms. (video *Huge Roman Catapulta* available on YouTube). Again, prestretched polyester was used. It could well be that these modern results with prestretched polyester come close to those achieved by Roman armies with horsehair.

There are rumours of far greater distances achieved in modern times, but they are best treated with caution as similar to those tales of the fish that got away.

There are other points to be noted about the trajectory and accuracy of a catapult bolt:

1. While it is possible to achieve a very tight 'group' on a target at point blank range, as in Figure 28, Roman bolts being hand-carved inevitably varied in shape and weight and so eventually separated and fanned out over long distances. The only surviving group of wooden shafts found to date were in the collapsed Tower 19 at Dura Europos. Not one is identical to another, there being particular variation in the shape of their flights (Baur and Rostovtzeff 1931, 72-3 and plate IX). The only Roman evidence as to how far bolts depart from their original line of flight is the missile spread on the Hod Hill 'target' of Huts 36a and 37, analysed in Figure 22 above. At an estimated range of 130 metres eleven bolts struck the ten-metre diameter Hut 37. Four bolts are close together next to the doorway in a four metre group. Are these the work of one marksman?

2. Differing figures have been suggested for the maximum range for effective shooting, which will be limited by missile visibility, and be affected by the wide-spreading path of bolts beyond that point. Schramm suggests 150m (Schramm 1918/1980, 25), Paul Holder 200m (Holder 1987, 3). Certainly, beyond 200m it becomes more and more difficult to spot the exact point of impact and so to be able to apply corrections of elevation and direction to the aim. The Vitruvian wood frame three-span and larger sizes of bolt-shooters must have been difficult to aim when a charging enemy was very close, because their solid frames blocked off the line of aim along the missile, or along the case, as advised by Heron (see Figure 61).

Figure 29 - Bolt leaving a Xanten-Wardt scorpion
A still frame from a video taken at the Tom Feeley Memorial Shoot catches the bolt leaving a Xanten-Wardt scorpion (photograph: the author).

Figure 29 demonstrates the excellent line of sight for aiming afforded by this small size of catapult; the artilleryman can look over the top of the spring-frame and use the top of the two rope springs to judge windage. By operating the scorpion from a kneeling position he presents the enemy with a much smaller target. The photograph also records the front recoil of the Vitruvian curved arms and the fact that they bring the bowstring right up to the centre stanchion, as evidenced by the damage on the Xanten-Wardt frame in Figure 37 left.

In contrast, it will be pointed out (Chapter 6 page 78) that the arch strut catapults' open frame allowed a 'wide screen' view of the enemy and unobstructed aiming along the missile and case, with the semicircular arch itself as an extra aid to framing a target. Soon after we fielded our version of the *manuballista* at public displays, two novice *ballistarii*, shooting at a plywood board 1.5m high and 1m across at a range of 60m, averaged one hit in every three shots. One of them achieved a 'Robin Hood' result: his first bolt pierced the target, his third gouged a groove along the shaft of the first.

The Bolt-Shooter: Accuracy, Range and Effects

Examination of the missiles that have penetrated the target in Figure 52 shows that their flights are in differing positions, proving that they have been spinning in flight. Aircraft engineer Don Acklam assures me that even a tiny error of a minute of a degree in the alignment of the flights on the bolt would cause this. Whether the spinning increases the accuracy of the trajectory, as with a bullet or shell, has not been tested. It may reduce the chance of the bolt cartwheeling.

In any case, the purpose of Roman artillery was to launch a dense swarm of missiles over distances well beyond the range of enemy counter-fire with arrows. The large arch strut bolt-shooters in good tune could send their bolts '...*not less than twice the range of a shot from a bow*', probably to at least 500m.

Bolts and arrows are one strike missiles. We estimate that the first bounce range of the smaller sizes of stone-throwers was between 150 and 180m; on hard ground these multi-strike missiles would continue to bounce well beyond that range, often fragmenting into shrapnel.

Roman firepower, as is frequently stressed in this book, was adjusted to be proportionate to the strength and threat of the opposition; at its most powerful it involved the simultaneous use of every type of missile launched by hand or machine. For an attempt to describe the ranges achieved by each type of missile and how a charging enemy would become enveloped in an increasingly dense hailstorm of death, see Arrian's detailed battle plan in Chapter 11 page 145 below.

Chapter 5

Reconstructing the Roman bolt-shooter

It is possible today to reconstruct the bolt-shooting catapult used by late Republican and early Imperial armies with a high degree of accuracy and authenticity because of the discovery at Xanten-Wardt in north west Germany of a bolt-shooter's spring-frame preserved with its wood, iron and bronze fittings, part of its stock, and even traces of the sinew-rope. This find, (Schalles 2010), provides by far the most important archaeological evidence to date, being so complete that it is no longer necessary to estimate the design and measurements of some components: the Xanten-Wardt *scorpio minor* spring-frame can be reconstructed with millimetre accuracy and be set alongside the list of bolt-shooter components and measurements by Augustus' catapult engineer Vitruvius, the illustrations on relief carvings such as the Vedennius' tombstone (Figure 35), and other finds of iron and bronze catapult parts (Figures 32 and 33).

The problems of reconstructing Greek and Roman catapults from the Greek and Latin artillery texts were eventually solved over a very long period by French, German and British pioneers. Unfortunately, the manuscript copies of the Latin text of Vitruvius' *De architectura* have suffered corruption and losses; furthermore, his diagrams have not survived. However, the distinguished British architect William Newton, whose father was Sir Isaac Newton's first cousin, not only produced the first English translation of all of Vitruvius' Ten Books (Newton 1791, published posthumously by his brother), but suggested possible replacements for Vitruvius' lost illustrations (Figure 54). He had already published in 1780 diagram reconstructions of the military machines described in Book X, recording it in French in order to reach the many continental scholars (Newton 1780). His English translation of the catapults in Book X, repeated his 1780 illustrations and added detailed and highly perceptive page notes emphasising the problems of understanding what these machines were like.

In the nineteenth century the challenge was taken up by Swiss engineer and army officer Guillaume-Henri Dufour (Dufour 1840). German artillery officer Colonel Köchly and classical scholar Professor Rüstow suggested several important improvements in the interpretation of the ancient texts (Köchly and Rüstow 1853). Emperor Napoleon III, himself an artillery officer and published author, encouraged Generals Dufour and de Reffye to make the first working full-size reconstructions of Roman catapults. They were particularly successful with their Vitruvian bolt-shooter and the *onager*. By then a scholarly Franco-German catapult war was under way. All researchers benefitted from an outstanding scholarly edition of the Greek texts on sieges and artillery, *Poliorcétique des Grecs*, published in 1867 by Carle Wescher, librarian at Napoleon III's

Imperial Library in Paris. It is still the standard work on the subject and is heavily relied on by all writers today. All the authors' texts in it are still quoted by Wescher's pages.

Understanding of the Roman bolt-shooter was greatly improved by Erwin Schramm, who was able to make use of the parts of a bolt-shooter's iron framework discovered at Ampurias in Spain in 1914, the first major parts of a Roman catapult to come to light (Figure 31). Schramm's 30-year programme of impressive reconstructions at the beginning of the twentieth century were backed by another imperialist, Kaiser Wilhelm II (Figure 30). There have been further discoveries of parts of Roman bolt-shooters during the 20th century, culminating in the 1999 discovery of the nearly complete Xanten-Wardt front frame.

Two major unsolved problems remained: first, the design of the stone-throwing *ballista* described by Vitruvius and by Heron of Alexandria. The version by Reffye and

Figure 30 - Schramm demonstrating his ballista *to the Kaiser (photograph: Saalburg Museum)*
In Figure 30 Major Schramm (centre) is demonstrating his reconstruction of Vitruvius' *ballista* to Kaiser Wilhelm II at the Saalburg fort on 16 June 1904. The Kaiser was almost a twentieth century victim of the *ballista* when the bowstring slipped under the missile and lifted it vertically (pers. comm. Dietwulf Baatz). He was moved quickly out of its descending path. (see Bowstring problems in Chapter 7).

Dufour was unsuccessful; Schramm's reconstruction of 1918 involved changing 20 of the 49 numerals in Vitruvius' text. Eric Marsden produced a preliminary model of the *ballista* (Figure 103), but his tragic early death prevented his intended further revision of the Latin text and production of a full-size machine.

The second unsolved problem was the five-page illustrated Greek text entitled 'Construction and Dimensions of the *Cheiroballistra*'. The text and illustrations of the *Cheiroballistra* (Latin *manuballista*. 'hand catapult') had been published in 1693 by Louis XIV's librarians from a single manuscript copy in the Paris Library. It has been by far the most difficult of all the surviving Greek catapult texts to understand, and its interpretation remains controversial today. The story of early attempts to make sense of the manuscript is given in Chapter 6 page 69.

The standard of artillery research was raised to a much higher level when in 1969 and 1971 Eric Marsden published his masterly study *Greek and Roman Artillery*, setting the subject in its historical framework and providing an excellent edition of the most important Greek texts. Since 1971 many artillery parts have been identified and meticulously catalogued by Dietwulf Baatz, but Marsden's early death prevented him from incorporating them in revised reconstructions of the Roman machines. I have attempted to keep up the momentum of his research, and the reconstructions described in this book are based firmly on a fresh examination of the texts of the manuscripts of the artillery writers and the archaeological evidence. Field testing of these reconstructions has enabled some assessment to be made of the original weapons' characteristics. Unfortunately, we have been unable to match the ranges and strike-power that were achieved with sinew-rope springs.

Augustus' creation of Rome's first regular professional army meant that for the first time there would be a standing pool of catapults available for each legion. The Tenth Book of engineer-architect Vitruvius' *De Architectura*, published c.25 BC, contains details and measurements of the bolt-shooting *scorpio* and stone-throwing *ballista*. He thanks Augustus for appointing him (I, preface 2) saying that, '*along with Marcus Aurelius, Publius Minidius and Gnaeus Cornelius I was put in charge of the construction and repair of* ballistae, scorpiones *and the rest of the* tormenta.' (Could 'the rest of the tormenta' – torsion catapults - include the one-arm *onager*? See the discussion on page 179). His remark that he had been '*first known to your father*', implies that he had also served Caesar as an engineer, perhaps also as a specialist in artillery.

The survival of Vitruvius' descriptions of *scorpio* and *ballista* has completely transformed our understanding of these machines, and provides the vital authentic background for identifying the increasing number of archaeological finds of catapult parts.

Archaeological evidence for the design of the Roman torsion bolt-shooter

The only major find known in Schramm's day was part of the corroded iron plating from the spring-frame of a three-span *scorpio* of the late second to early first century BC. This was found in 1912 in an arsenal near the South Gate of the Greek city of Emporion, present-day Ampurias (Spain), complete with its four bronze washers. The Ampurias spring diameter is 79 mm. Unfortunately, the plating of the central stanchion was missing. Professor Jacobi of the Saalburg Museum identified the iron framework as from a catapult of the type already reconstructed by Schramm and kept at that museum. At the direction of the German Emperor, Schramm visited the Barcelona Museum to study the frame and some missiles. His excellent reconstruction (Figure 31) was published and discussed by him at length (Schramm 1918/1980, 40-46). An important and unique feature of this frame, as reconstructed by Schramm from Spanish experts' drawings and photos, is the flat front of the side-stanchions: all frame parts discovered since then and the Vedennius relief display side-stanchions with the protruding curved profile recommended by the artillery texts to compensate for the weakness caused by the deep cut-outs at the rear of the side-stanchions to accommodate the forward travel of the arms.

The curious case of the Ampurias frame

Information from Martin McAree and a photograph by Salvador Busquets (Figure 32) drew our attention to a surprising and important correction to the Spanish and German publication of this frame, revealed by the remounting of the display in the Barcelona Museum; the frame is now set upright rather than flat on its face.

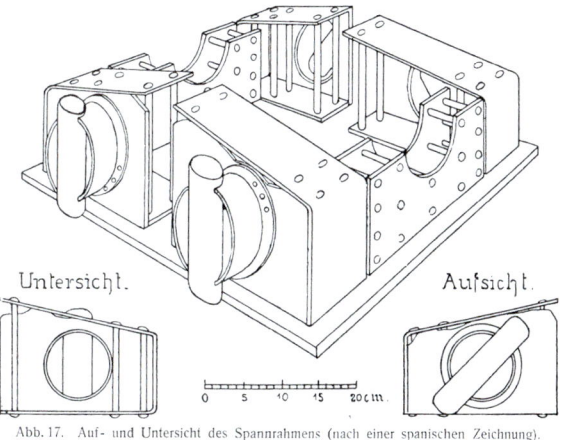

Figure 31 - The Ampurias frame as published
(left) Schramm's fine reconstruction at the Saalburgmuseum of the bolt-shooter discovered at Ampurias in Spain in 1912 (photograph: the author). **(right)** Schramm's Abb. 17 based on a Spanish drawing.

Figure 32 - The corrected Ampurias display
(left) The frame mounted vertically, revealing standard curved front side-stanchions (photograph: Salvador Busquets). **(right)** Len Morgan's reconstruction, with a suggested restoration of the missing plating of the centre-stanchion based on that surviving on the Caminreal frame. (photograph: Len Morgan).

The Caminreal iron frame plating

In 1983 the complete but crushed iron plating and bronze washers from a slightly larger *scorpio* frame, dating to c.80 BC, (Vicente et al. 1997, 167-199) were found in a villa at Caminreal (Teruel), Spain (Figure 33). Its washer spring diameter of 84mm would have made it a 3 1/3 span; but the holes in the frame are only 80mm in diameter. Therefore, the frame was originally for use with washers of about 74mm internal diameter and three-span size. Is this the result of an urgent battlefield repair having to be made at a time of shortage of spare washers of the correct size? It dates to *c.* 80 BC and the Sertorian War. The pattern of the nail holes in the plating of the centre stanchion prove that it uses a single centre stanchion and not Philon's two-part one, and is the uprated design described by Vitruvius. This is one of the key differences noted by Marsden in his analysis of the changes from Philon's to Vitruvius' catapults (1969, Appendix 1). It is therefore certain that the design of scorpion that Vitruvius

 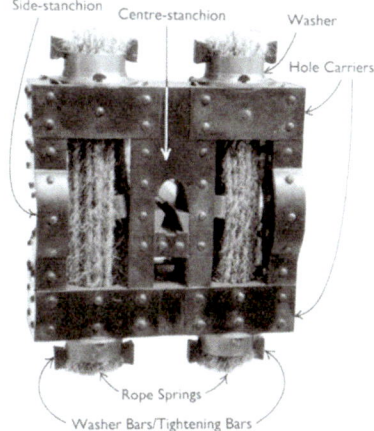

Figure 33 - The Caminreal (Teruel) iron frame reconstructed
(left) Reconstruction of the Caminreal scorpion by Len Morgan, in action at Piddington catapult trials. The design and proportions of the arms, case (stock) and stand are based on Vitruvius' text. (photograph: Margery Wilkins). **(right)** Reconstruction of the Caminreal frame in the museum at Aalen (photograph: Dietwulf Baatz). The square cutout at the bottom of the Caminreal's centre plating is interpreted by the author as housing a wedge system clamping the case to the frame.

describes was in service by the time of the Sertorian War. Agesistratos may have been involved in its development; Marsden's conclusion (1969, 206) was that he was at his peak about the second quarter of the first century BC.

Catapult Washers

There have been numerous finds of Hellenistic and Roman catapult washers, including some from the shipwreck at Mahdia in Tunisia of the second quarter of the first century BC (Figure 34). See Baatz 1985. Moulds for the lost wax casting of bronze washers have been found at Auerberg, Germany (Handlesbank 1989).

The Xanten-Wardt smaller scorpion (*scorpio minor*)

The previous section summarises the very limited archaeological evidence afforded by the Ampurias and Caminreal outer plating and other finds. No wood from a spring-frame had been found, although estimates of it could be made from Vitruvius' description and the surviving frame plating. The Xanten-Wardt frame from a Roman torsion bolt-shooting catapult of the mid first century AD was discovered in 1999 in north west Germany, in the gravel bed of a former loop of the River Rhine, close to the Xanten Roman Archaeological Park (Figure 36). Most of the sumptuous official report is rightly devoted to the details of the painstaking recovery of the machine

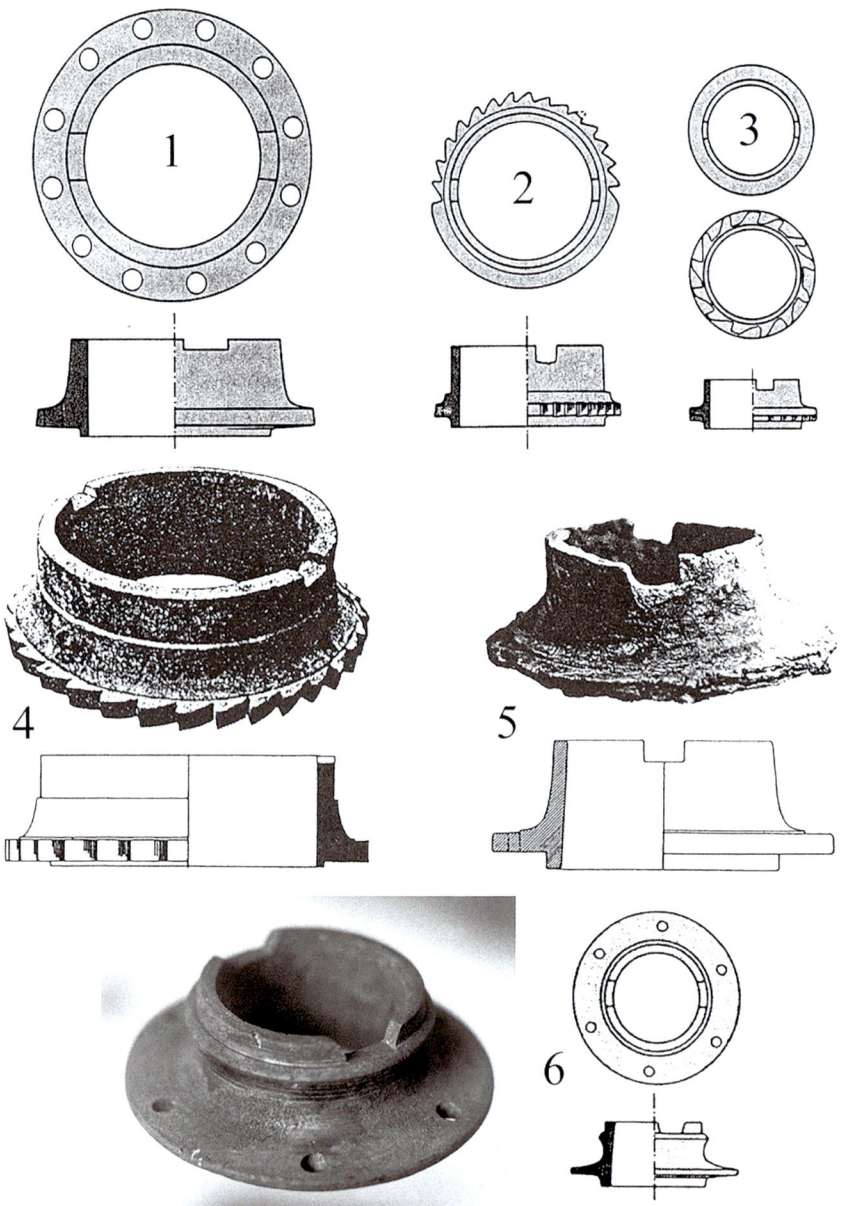

Figure 34 - Catapult washers (diagrams after Dietwulf Baatz. Washer photograph: the author)
The washers in Figure 34 are presumed to be from bolt-shooters, from various sites and periods. **1**, **2**, and **3** are from the shipwreck at Mahdia (Tunisia) dated to the second quarter of the first century BC. Like number 4, numbers 2 and 3 use ratchet tracks to lock the washers into position, instead of the usual system of pins and holes. **4**. Sunion (Greece). third century BC. **5**. Azaila (Spain) Dated to the Sertorian Rebellion of 80 – 72 BC. **6**. Bath (England) Inner diameter 4.1 cm. This bronze washer had been deposited in the sacred spring, possibly by an artillery engineer as an offering to Sulis-Minerva. Note the lathe finishing marks. It dates from the second half of the 1st century or first half of the second century AD.

from its 'coffin' of solidified sand, grit and pebbles. The anaerobic conditions created by this 'coffin' preserved the European ash frame, the bronze and iron plating, the bronze washers, iron washer-bars and one complete washer pin, and even a trace of the sinew-rope itself. One major loss, of the spring-frame's left-hand side-stanchion (Figure 37 below left) is actually a bonus which allows a clear view of inner details like the iron plates and rivets used to secure the case to the centre stanchions; otherwise the frame parts are complete. Unfortunately, only the front section of the wooden case and slider survive, the rest having been sawn off and discarded in Roman times (Figure 37 right).

The case of the Xanten-Wardt catapult was permanently fixed to the spring-frame, being clamped to it by four long bolts through the centre-stanchions. The Roman engineers confronted a major problem because the case had somehow come adrift, perhaps splitting along the line of these bolts. An attempt had been made to refit one, but the continued existence of the two clamping bolts made this impossible: in order to allow the end of a new case to be inserted they would have had to be removed, which would have involved dismantling the frame to enable new bolts to be driven through the centre-stanchion and case, and their heads to be peened. This would have required several man-hours work in the *fabrica* (workshop), and may have been the reason why the damaged frame was discarded. It is surprising that the washers and other parts were not rescued. Perhaps its demise was accidental and it fell overboard from a ship before it could be sent to the *fabrica* for a rebuild.

The machine bears a striking resemblance to the detailed bas-relief of a bolt-shooting catapult on the tombstone in the Vatican Museum of C. Vedennius Moderatus, who was appointed by Vespasian and Domitian as *arcitectus armamentarii imperatoris* (Engineer of the Emperor's Arsenal) (Figure 35). Heavy plating is riveted onto the front of the hole-carriers. Note the incised circles on the side of the side-stanchion, representing the rivet heads found on the Ampurias, Caminreal and Xanten-Wardt frames (Figures 31, 33 and 36). Also depicted are the washers, the ends of the washer bars, the rope springs, the four eye-pins locking the washers, the bowstring passing round the arms, and what appears to be the point of a bolt in the aperture of the battle-shield. The only major difference is that the Vedennius relief has an all-in-one battle-shield like that found at Cremona (Figure 45); whereas the Xanten-Wardt has two separate plates protecting the rope-springs. The relief is believed to date to *c.* AD 100.

The Xanten-Wardt frame still has one complete example of the eye-pins locking the washers, a type of pin previously known from the Vedennius relief. An apparently identical pin was handed in to Cumbria and Lancashire Portable Antiquities Scheme Finds Liaison Officer Dot Boughton in 2013, found by a metal detectorist on a site in Cumbria. Reference LANCUM –F01835 PIN. The catapult is extremely elegant and superbly finished, with the side and front bronze plating given a neat edging, a

35. Vedennius' scorpio and the Xanten-Wardt reconstruction
(left) The detailed relief of a *scorpio* on the side of the tombstone of artillery engineer C. Vedennius Moderatus, dated to *c*. AD 100, in the Chiaramonti Gallery of the Vatican Museum. **(right)** The reconstructed Xanten-Wardt frame from the same angle. (photographs: the author).

Figure 36 - Xanten-Wardt frame
The Xanten-Wardt scorpion frame after 22 months of painstaking recovery and conservation from a block of concretions by Jo Kempkens and Ton Lupak (photographs: Maarten Dolmans).

purely decorative feature. It is a reminder that far from being roughly finished, Greek and Roman catapults were made to look impressive, as Philon urges (*Bel.* 61, 29; 66, 19). It is further proof of the Roman soldiers' great pride in the appearance of their equipment.

The identity of the Xanten-Wardt catapult

The description of the machine as a *manuballista* (hand catapult) in the title of the 2010 report and by some of its contributors is incorrect, because at the time of the Xanten-Wardt machine the word *ballista* was only used for stone-throwers. The term *manuballista* belongs to the later design of bolt-shooter with a metal frame, of the type described in the *Cheiroballistra* manuscript (Chapter 6 page 69) and first known from its appearance on Trajan's Column in the Dacian campaign of AD 101-102. The striking change of terminology that applied *ballista* to the metal frame bolt-shooters, as *manuballista* and *carroballista*, can be explained by the fact that they had borrowed the special palintone offset framework that gave the *ballista* stone-thrower its increased power.

The Xanten-Wardt machine is a bolt-shooting *scorpio minor* powered by torsion springs of sinew rope, which were mounted in a hardwood frame of European ash reinforced with bronze and iron plates. While the term *scorpio* was sometimes used in a general sense as an alternative to *catapulta*, it was frequently applied to the smaller sizes of bolt-shooting *catapultae*. The distinction which Hellenistic and Republican writers made between the larger sizes of bolt-shooters and scorpions is maintained by Vitruvius, who talks about the tuning of *catapultae*, *scorpiones* and *ballistae* (Vitruvius I.1.8). The historian Livy lists *ballistae* and *catapultae* amongst the booty taken by Scipio Africanus in Hannibal's base at Cartagena (See page 14); he also records '*a vast number of larger and smaller scorpions and armour and weapons*'. In his eyewitness account of the sieges of Jotopata and Jerusalem, Josephus uses a pair of Greek words for bolt-shooters, *katapeltai* (= *catapultae*) and *oxybeleis* ('sharp-firers' = *scorpions*). There is evidence from Justus Lipsius, Sisenna and others, quoted by Marsden (1969, 89) to identify the three-span bolt-shooter as the *scorpio maior* (larger scorpion), and the one-cubit (two-span) as the *scorpio minor* (smaller scorpion). The term *scorpio* was probably used for the smaller sizes of *catapultae* because, like the insect, they were small in size but deadly. In the late Roman Empire the one-arm *onager* stone-thrower was nicknamed a scorpion; Ammianus Marcellinus (XIX, 4, 4) explains that this was because of the visual resemblance of its single vertical arm and sling to the upright claw of the insect.

The clearest evidence to confirm that the Xanten-Wardt machine is one of the group of scorpion catapults, as a *scorpio minor*, is found in Philon's detailed description of Dionysius of Alexandria's winched, stand-mounted, torsion repeating catapult (*Bel.* 73, 24-6, Marsden 1971, 146-7) (see my Figure 5). He calls it a *skorpidion*, a small

Figure 37 - Xanten-Wardt missing parts (photographs: Maarten Dolmans)
In Figure 37 (left) the missing left-hand side-stanchion reveals the rectangular plates and long bolts (white arrows mark the top pair) which lock together the components of the centre-stanchion. The indentations on the back of the wooden centre stanchions (red arrows) were caused by the impact of the bowstring, confirming that the missing arms were curved, as described by Vitruvius, in order to increase the forward travel of the bowstring. Side view (right), with the sawn-off remains of a replacement slider and case. See the discussion above as to why this case and slider could not be fitted.

scorpion, '*not much larger than a One-cubit machine, and not much smaller than a three-span.*' Both sizes are by implication winched, stand-mounted scorpions (although Dionysius' machine had a unique double chain drive instead of rope to achieve pull-back). A three-span scorpion had a spring-hole diameter of 4 Greek dactyls = 77mm and a bolt 36 dactyls = 69cm long. A one-cubit or two-span machine had a spring-hole of 2⅔ dactyls = 51mm diameter, and a bolt 24 dactyls = 46cm long. Philon states that Dionysius' bolt was only one dactyl longer than that of the one-cubit machine at 25 dactyls = 48cm long. In comparison the Xanten-Wardt torsion catapult has a spring-hole diameter of 2⅓ dactyls = 45mm, and a bolt 9 x 45 mm = 40.5cm long. This makes it slightly smaller than a one-cubit/two-span, as a one-and-nine-tenths-span.

Unfortunately, Schalles' Xanten-Wardt report ends by Baatz identifying the catapult as hand-held and hand-spanned. The report also calls the catapult a *manuballista*, the name of the later arch strut metal frame bolt-shooter. This interpretation of smaller torsion catapults as stomach-bows lacking a winch and stand was sparked off by Dietwulf Baatz in his 1974 *Saalburg Jahrbuch* article on the finds from Gornea and Orşova, and repeated in his 1978 article in *Britannia*. It contradicts Heron of Alexandria's authoritative statement, in his description of the development of artillery (*Bel*. 84-5), that torsion catapults developed so much power that the stomach-bow's withdrawal-rest had to be replaced with a winch, with a pulley-system added for the larger machines. Baatz later told me in a letter that his interpretation is based on the theoretical possibility that a Roman engineer could have picked up the idea of loading by stomach pressure from Heron's description of the early Greek stomach-bow, and applied it to a small torsion catapult. Christian Miks published a reconstruction which follows Baatz' interpretation (Miks 2001, 153 – 183).

Irrespective of the evidence reviewed above for the *scorpio minor* torsion catapult as winched and stand-mounted, the complete survival of the wood and metal of the Xanten-Wardt frame enables fully accurate replicas to be made, and settles the controversy. There are now several copies of our Xanten-Wardt reconstruction in use by re-enactment groups in Britain, and others in Ireland and Holland. Tom Feeley's version is in the Carvoran Roman Army Museum on Hadrian's Wall, and Len Morgan and my version will eventually be on permanent display at Comlongon Castle near Dumfries. We loaned one to the British Museum for their 2024 exhibition *Legion: Life in the Roman Army*. They have been made with millimetre accuracy by Len Morgan and Tom Feeley from the official plans, and with help, through Maarten Dolmans, from Alexander Zimmerman and Xanten-Wardt restorer Jo Kempkens. We have restored the arms, winch and stand from Vitruvius' description and measurements. The simple, critical test is to discard the support stand, remove our winch and then attempt to hold the weapon steadily in the aim; next try loading it by stomach or muscle power, inserting the bolt, lifting it back into the aim, all this time and time again as required on the battlefield. To do this for any length of time without a support stand is impossible. The point of balance is immediately behind the frame.

In any case the matter is settled by the fact that rope springs of 45mm diameter are far too powerful to be armed by stomach and muscle power. We proved this when testing our reconstruction of the *cheiroballistra* with its identical spring size. The reason I used side ratchets on that reconstruction was to allow attempts to arm it as if it were a stomach-bow. Such attempts failed, having moved the bowstring back by no more than a few centimetres. When Len Morgan tested the force required to pull the arms fully back, he found it to be 335 kg. The extra power available from the final 5.5 centimetres of drawback and the resultant gain from the extra degree or so of arm travel was neatly proved by the fact that a 27.4 kg load was required to achieve this final small distance.

Figure 38 - Operating Xanten-Wardt scorpions
Eleanor and Tom Elder operating the first Xanten-Wardt reconstructions by Morgan and Feeley, for David Sim's penetration tests. The Vitruvian curved arms were not then available. At a range of 40 metres these catapults punched square holes in the replica iron shield bosses (Figure 95). (photographs: the author).

With a stand and levered winch, operators of whatever stature and muscle power can keep reloading and aiming for indefinite periods. The use of the support stand meant that two hours of shooting trials of the Xanten-Wardt reconstruction could be carried out by my two friends Tom and Eleanor Elder, then aged twelve and ten, who put on a display of accurate shooting (Figure 38). Even the Elginhaugh size of scorpion with its 34mm washer diameter, at present under reconstruction by Len Morgan and the author, is too powerful to be spanned by stomach pressure.

The above lines of argument and the evidence from Philon for the Xanten-Wardt torsion catapult and those of similar calibre as smaller scorpions with winches and stands do not appear to be given space in the Xanten-Wardt report. The end of the report presents the reader with a curious reconstruction as a hand-held catapult loaded by pressing the extended slider vertically down into the ground, apparently using the weight of the machine plus downward pressure by the operator who is shown pulling back the bowstring with his hands. The illustration of it being aimed (Figure 39) shows a reenactor holding the stock with

Figure 39 - The Xanten-Wardt Report's box support (after Schalles 2010, 175 Figure 5)

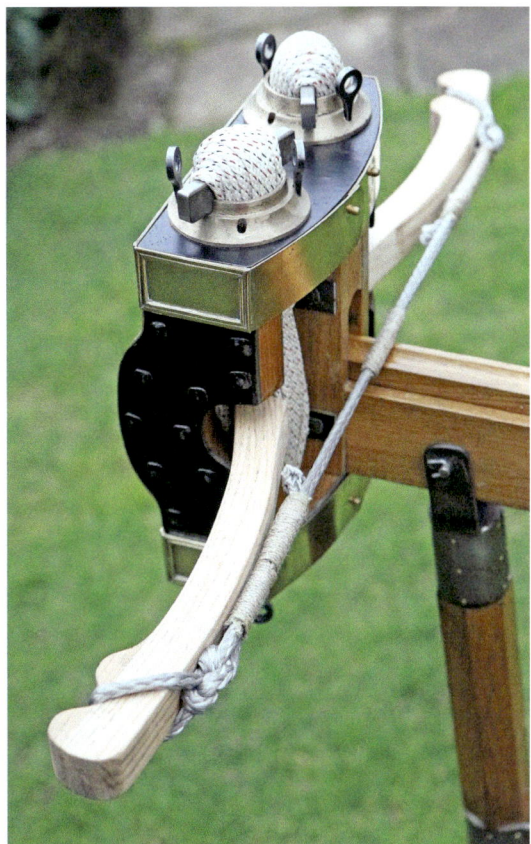

Figure 40 - Reconstruction of the Xanten-Wardt scorpion
(left) The case, slider, trigger, windlass box, arms and stand from the details supplied by Vitruvius. **(right)** The author's curved arms, following Vitruvius' measurements. (photographs: the author).

his left hand, with a curved shoulder-piece resting against his right shoulder; he has to take the weight of the weapon by raising his left leg onto a large box, and supporting his left elbow on his left knee. This strange box support system confirms that there is a weight problem. We estimate the pull back as at least 275 kg., impossible to achieve by stomach or muscle power. The vertically mounted shoulder-piece cannot be used to apply stomach pressure.

The size formula for bolt-shooters

The formula established by the Greek engineers was that all the bolt-shooter's components are in fractions or multiplications of the spring-hole diameter, which is one ninth of the length of the bolt. The internal diameter of the spring-hole in the washers decided the maximum diameter of the spring-rope that could be crammed in, and can be regarded as the calibre of the catapult. The Xanten-Wardt *scorpio minor* adds to the growing evidence that the 45mm calibre was a popular size of bolt-shooter, because other washers of this size have been found, one from the Tunisian Mahdia shipwreck (Mahdia no. 3, 45mm), and another from Volubilis in Mauretania (no. 467, 44mm). Furthermore, during the Carlisle Castle Green Millenium Excavations of 1998-2001 two blocks of iron-bound ash were found in the demolition layer of a military workshop or store dated to *c.* AD 140 (Howard-Davis 2009, 713 Figure. 364 and Plate 213). Their suggested identification by Mike Bishop (Howard-Davis 2009, 712) as parts of a catapult was confirmed when Tim Padley, Keeper of Archaeology at Tullie House Museum, kindly allowed me to place one of them next to the Xanten-Wardt reconstruction and the official 1:1 plan of the German find (Figure 41). They are the same in size and outline, the two Carlisle hole-carriers

Figure 41 - Carlisle catapult find
Iron-bound ash block 27, one of two from the Carlisle Millennium Project, next to the 1:1 plan of the Xanten-Wardt hole-carrier (photograph: the author, by kind permission of Tullie House Museum, Carlisle).

being in the early stages of construction, with the spring-holes, almost certainly of around 45mm diameter, and other details yet to be cut. The outer dimensions of the Xanten-Wardt hole-carrier are 208 x 87 mm. This is the first hard confirmation that standardised catapult plans were distributed to Roman workshops (*fabricae*) throughout the empire. The Carlisle objects are the first parts of the framework of a catapult to be identified in Britain; up to now the only British artillery finds, apart from the hundreds of boltheads, have been the washer from the spring at Bath (Figure 34 above, 41 mm diameter) and the one from Elginhaugh fort, Scotland (34 mm). 21/3 dactyls, 45mm, is also the emended reading suggested by Marsden for the spring-hole diameter in the *Cheiroballistra* ms (Chapter 6 page 96).

Constructing the spring-frame

Our reconstruction follows my revised edition of Vitruvius' text (Wilkins 2000). In the following discussion of Vitruvius' instructions the Latin word for spring-hole, *foramen*, is translated as 'hole', h.. Vitruvius' square six h. high x six h. wide spring-frame outer dimensions may have been the textbook standard. He says that the frame can be made higher (*anatonus* = 'oversprung'), or lower (*catatonus* = 'undersprung'); in both cases the length of the arms must be altered. None of the frames found so far are square: the Ampurias, Caminreal and Cremona finds are all from spring-frames, ¾ h., 1 h. and 1/3h. undersprung respectively. The Xanten-Wardt frame is the same height, 4.9 h., as the Ampurias and Caminreal frames, but is only 4.6 h. wide which makes it oversprung. Vitruvius (X, x, 6) says, '*if the frames are ... oversprung the arms will be shortened, so that, to compensate for the weaker spring caused by the height of the frame, the shortness of the arm may deliver a more violent blow.*' This narrowing of the Xanten-Wardt frame meant that our modern rope-springs tend to rub on the frame; not a problem, perhaps, with the slippery oil-soaked sinew rope. The reason for this narrowing may have been to save timber when constructing large batches of catapults.

Obviously, the tensioning and twisting of the two rope springs applied enormous forces of compression and torsion to the frame in which they were set. Before the invention of artillery, Greek engineers had used hardwood frames encased in metal plates for heavy duty machinery such as cranes. These were the materials used for all catapult frames until the introduction of the Roman army's arch strut catapult with all-metal framework at the end of the first century AD (Chapter 6). The constructional techniques required were within the skills of any competent carpenter, blacksmith and bronze worker, and well within those of the specialist engineers in a Roman legion's workshop. Catapults were constructed, repaired and operated by the legions, although some auxiliary regiments may have been allowed to operate stone-throwers from the second century AD. An important papyrus from Egypt (Bishop and Coulston 2006, 42, Figure 17) records two days production in a legionary workshop making catapult frames and a variety of arms and armour, using a supplementary in-house unskilled workforce which included auxiliary soldiers, soldiers' slaves, *immunes* and civilians, the latter possibly including the families of the garrison. Wives and daughters would have had the skill needed to spin the sinew fibres into rope. Casting round bronze components like catapult washers and finishing them on lathes was a long-established procedure; lathe marks can be seen on several examples of washers, for example on the Bath washer (Figure 34 above). The two Carlisle unfinished frame parts are important evidence of catapult production on Hadrian's frontier.

The plating covering the spring-frame

> 'Let the four anguli qui sunt circa *(outside angles)* on the sides and faces be bound with iron plates and bronze bolts or nails.' (Vitruvius, X. 10.3)

Even before the discovery of the Ampurias plating in 1912, the existence of plating was confirmed by the Vedennius relief (Figure 35). The plating from the front of the centre-stanchion is missing from the Ampurias find, but has survived on the Caminreal frame, as have all the nails (Vicente J.D. *et al* 1997). On the Ampurias frame the bronze washers sit on large iron counter-plates which eliminate wear on the wooden hole-carriers. The Caminreal frame has four counter-plates in the form of washers, pinned against rotation by four projecting tabs. The Xanten-Wardt has no counter-plates; the bronze washers sit directly on the iron plating of the hole-carriers. There are iron plates, thickened in their centre on the vertical sides of the two outer stanchions, and bronze plates elsewhere (Figure 35 right and Figure 37).

John Anstee claimed (pers. comm.) that plating was unnecessary, and that it was only there to protect the wood against the fire-brands and arrows which were often used to destroy these highly inflammable machines. However, Heron (*Bel.* 92-3), describing the stone-thrower's stanchions, says that plates must be nailed to both sides of a stanchion, and plates encasing its double tenons must be fixed with nails '*so that the side-stanchion, bound securely on all sides, may be able to withstand the strain*'. Plating the tenons, which are sunk out of sight in the mortise slots of the hole-carriers, can have nothing to do with fire-proofing.

Importing metal plates

David Sim has made the remarkable discovery (Sim 2000, 1-3) that the Roman army was using high quality <u>rolled</u> iron plates of uniform thickness (variation of no more than 0.1mm). These may well have been imported from India along the busy caravan routes that passed through Syria, or by sea through the Arabian Gulf to Egypt. Known today as wootz steel, Indian steel with a high carbon content is recorded as being produced there from as early as the 4th century BC. Alexander the Great is reported to have been given 100 talents (2,620 kg) of Indian iron (Quintus Curtius IX, viii, 1). Several Greek and Roman sources describe the high reputation of Indian iron: the *Periplus of the Erythraean Sea* states categorically that it was being imported from there (Bronson 1986, 13-51).

Pliny (Nat. Hist xxxiv. 145) states that '*of all the types of iron that from the Seres takes the palm*' (*ex omnibus autem generibus palma Serico ferro* est). Commentators tend not to accept Pliny's Chinese origin, believing that the metal was manufactured in India and that the Romans had been persuaded to believe in a Chinese source by the traders who supplied it. This may be so. A recent paper (Srinivasan, Sinopoli and Morrison 2009) records iron and steel high carbon artefacts and crucibles dating from 800 to 400 BC from excavations at Kadebakale and elsewhere in Southern India.

However, in his next sentence Pliny adds that '*the Seres send this along with their clothes and skins*' (*Seres hoc cum vestibus suis pellibusque mittunt*). I think that he could be right, and that some part of this supply of rolled plates, or at least the technique of producing them, originated from China, being transported along the Silk Road in the same wagons as their silk fabrics and pelts. During the period of the Warring States, 403-221 BC, the Chinese were already using blast furnaces for large scale production of high-quality iron swords, spears, helmets and plates for armour.

David Sim has shown that rolled plates were used to produce the socketed boltheads for catapults (Figure 53). It would have been highly advantageous to use them for the catapult frame cladding: the ready prepared sheets would offer the Roman blacksmiths an alternative to forging the plates by long hours of hand hammering.

Figure 42 - Frame wedge system on a scorpio maior
A reconstruction by Len Morgan of the Vitruvian *scorpio maior*, showing the author's suggested wedge system clamping the case to the frame. The upper wedge is permanently fixed to the underside of the case, the lower one is hammered in from the front of the frame. This allows quick separation of the case and slider from the frame, solving the problem of the fixed all-in-one case and frame of the Xanten-Wardt *scorpio minor*. (photographs: the author)

Sim and Kaminski argue (2012, 54-57) that the Romans had the techniques and equipment that would have allowed them to produce machines with rollers to turn out the plates, but admit that even if these had ever existed they would be unlikely to have survived, or to be identifiable by modern archaeologists. The fact emphasised by Sim in his 2000 article is that Roman socketed catapult boltheads from sites all over the Empire are produced from plates that do not vary in thickness by more than +/- 0.1mm. It is difficult to avoid the conclusion that this was because they came from a single source, rather than being produced by Roman workshops all over the Empire.

The arms

> *The length of the bracchium (arm) is **7** h., its thickness at the bottom is 5/8 h., at the top 7/16 h.. Its curvatura (curvature) is 8 h.*' (Vitruvius X.10.6)

No wooden arm from any plated wood frame catapult has been identified. Philon has the same figure as Vitruvius for the length. Confirmation of the existence of arm curvature comes from the relief on the Pergamon altar (Figure 44). By *curvatura* Vitruvius means radius. Curved arms allowed the bowstring to end up closer to the spring-frame (Figures 40 and 44) and the arms to travel in a greater arc, thus storing a greater amount of energy. We have followed Vitruvius' figure for the radius in reconstructing the Xanten-Wardt's missing arms, but have not yet experimented with shortening the arms for this over-sprung design, as he suggests.

The case and the slider (i.e. the stock)

Vitruvius describes a laminated construction for the case, with a central beam and two side- or cheek-pieces, all 19 h. long (Figure 43).

> *The length of the canalis fundus (slider) is 16 h., its thickness <6/16> h., its width 1/2 h*

Canalis fundus literally means '*the base of the channel*', the channel being the groove along which the bolt travels. The case is 19 h. long and 1 h. wide and high. The Cremona battle-shield (Figure 45) supports a reading of ½ h. for the slider width and 6/16 h. for the thickness. This is a good example of an archaeological find providing a measurement where the figure in the manuscripts of Vitruvius is missing or corrupt. The extra 3 h. length of the case must be for the winch mechanism. The stump of the Xanten-Wardt slider has a taller profile and is 11/16 h. thick and 5/8 h. wide.

Only the front ends of the Xanten-Wardt slider and case have survived, the remainder having been sawn off immediately behind the frame (Figure 37 right). My interpretation of the square hole on the bottom edges of the larger Caminreal frame (Figure 33) and the Cremona battle-shield (Figure 45) is that this was a hole for a wedge driven in from the front to clamp the case to the frame, allowing the two to be separated quickly when the catapult was not in use (Figure 39 above). A companion wedge is permanently fixed to the underside of the case. This appears to be the solution on these larger catapults to the problem of the Xanten-Wardt's fixed, all-in-one construction. It allows the heavy three-span to be separated into two parts for easy carrying.

The trigger assembly

The general form of these components was worked out by Schramm, and was probably similar for bolt-shooters of all periods. So far, no catapult trigger parts have been identified.

The winch box

Vitruvius calls this a case (*camillum*) or box (*loculamentum*). In the author's reconstruction this structure does resemble three sides of a box. A similar design has been used for the Xanten-Wardt reconstruction, with Vitruvius' dovetailed back plate to lock the two sides rigidly together (Figure 40 left).

The stand and universal joint

Vitruvius' extraordinarily detailed list of parts for the stand contrasts sharply with the lack of detail given for more vital components. His text has lost some blocks of information in transmission: no mention survives of the all-important washers, for example.

Figure 43 - Scorpion case, slider and winch (photographs: the author)
In Figure 43 **(left)** are isometric cross-sections of the *canalis fundus* ('base of the channel' i.e. the slider), and the case, which is laminated from a central *canaliculus* ('channel') and two *regulae/bucculae* ('sidepieces'). The three pieces forming the sliding dovetail are marked in black, and can be cut from one piece of timber. **(centre)** The artilleryman's view of the winch box, drums, ratchets and a handspike, with the trigger mechanism and a three-span bolt loaded into the channel of the slider. Note the large ring spanner for turning the washers, on the ground to the left of the winch. **(right)** The winch, case, slider and trigger on our Xanten-Wardt reconstruction.

Figure 44 - Catapult on the Altar of Zeus, Pergamon. Scorpion curved arms (photograph: Margery Wilkins)

Figure 44 **left** is the stylised representation of the front of a bolt-shooting catapult on the Altar of Zeus from Pergamon (Turkey). It is dated to the reign of Eumenes II, 197 – 160/159 BC. **(right)** A battery of scorpions is shooting for David Sim's penetration tests: in the distance Tom Feeley is testing two *manuballistae* and two Xanten-Wardt reconstructions; in the foreground Len Morgan has just launched a three-span bolt from a *scorpio maior*. Note how the curved arms allow the bowstring extra travel, bringing it right up to the spring-frame.

The *caput columellae*, 'head of the column', is ingeniously translated by Marsden as 'universal joint'. This key component allows the case to swivel in all directions and gives the artilleryman total freedom to adjust the aim. In more recent times Robert Hook (1635 – 1703) is credited with inventing this type of joint. It was almost certainly used on Hellenistic cranes long before its application to torsion artillery.

The battle-shield (Figure 45) from a three-span bolt-shooter of the Fourth Macedonian Legion, was found with its washers in a military grave outside the walls of Cremona (Italy) along with human bones and two or three damaged skulls. These probably belonged to victims of the battle of AD 69 between the forces of Vitellius and Vespasian, recorded by Tacitus in Histories III, 29. The shield was first identified correctly by Mark Hassall and Peter Connolly (Hassall 1999, 24). The consular names date its manufacture to AD 45. The Cremona shield is described as 'a flat iron plate covered with a thin plating of copper'. This would make it strong enough to provide protection for the rope springs from enemy fire-arrows and other missiles, and partial protection from the weather. It also adds to the impressive look of the machine. Because the bolt is always positioned halfway up the spring-frame, it is clear that the shield has lost a section at the top. In this diagram the author has restored the shield's original outline and Claudius' imperial titles (he eschewed IMP). The probable position of slider and bolt is added in black. The rectangular hole at the base matches the lower hole on the Caminreal frame (Figure 33) and is interpreted by the author as allowing a wedge to be driven in from the front to lock the spring-frame to the case (Figure 42).

Figure 45 - The Cremona battle-shield

The standard bolts

The central hole in the Cremona shield provides evidence for the width and flights of a bolt. The shaft of the bolt must have had a diameter no greater than 18mm on a three-span *scorpio*, with the side flights limited to a projecting width of 8 mm. The aperture for the bolt on the Xanten-Wardt frame is 26 mm wide, allowing for a maximum bolt diameter of about 12mm plus flights. The Greek word for flight, *pteron,* means '*a feather*' or '*wing*', but also anything that is wing-like. Feather flights were probably standard for bolts until the wide-open frames of the arch strut bolt-shooters allowed the new design of bolt found at Dura Europos with wooden flights, as described by Procopius. See Chapter 6 for the terrifying, deadly penetration pattern of this bolt design. The first bolt for the Vitruvian type of catapult with its wooden shaft complete has been found in the bank of the Rhine in The Netherlands (Figures 3 and 47). The absence of wooden flights means that it had flights of feathers.

Roman Imperial Artillery

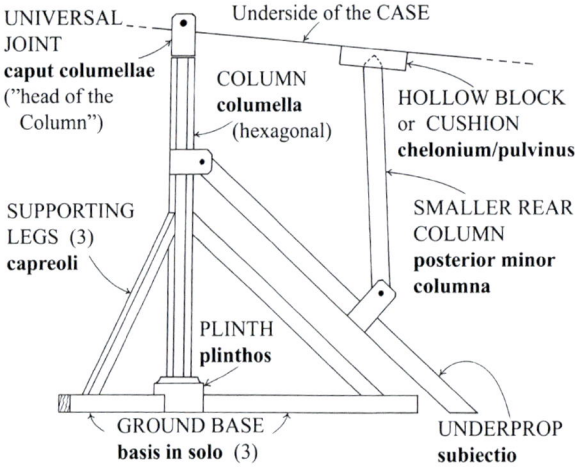

Figure 46 - Scorpion stand
(top) The three-span scorpion built by engineer Tom Feeley from the author's interpretation. (photograph: Tom Feeley) **(bottom)** Vitruvius' stand, drawn from the information in his text.

Hundreds of iron missile heads survive in museums all over the Roman Empire; many of them are too corroded to make estimates of their original weights feasible. Because of this it is not always easy to be certain as to which missile heads belong to arrows, or catapult bolts, or even javelins. However, one type of missile head (Figure 27) is generally agreed to be from a catapult bolt - the larger tapering, pyramidal bodkin point mounted on a hollow socket. This identification is confirmed by its use on the later Dura Europos type of catapult bolts, whose wooden shafts survive (Figure 63) and whose uncorroded heads can be weighed. The well-judged comments of Bill Manning (Manning 1985, 175-6) on the different designs of boltheads should be noted. A missile head with a socket, whether for a spear or catapult bolt, is designed to take the impact force of a heavy shaft and ensure that the head is driven fully home and does not snap off on impact. The internal diameter of this socket is critical to deciding the size of catapult that launched it, because of the above-mentioned size restriction of the exit hole in the centre stanchion.

A complete catapult bolt from the Rhine frontier

This is the first ever find of a standard Vitruvian, pre-*manuballista* bolt with its wooden shaft complete. It had been shot across the Rhine from the fort on the site of the modern Castellum Hoge Woerd at De Meern, near Utrecht, The Netherlands (Figure 47). Because it landed in the soft clay of the river bank it was preserved intact

Figure 47 - The complete number 211 bolt as conserved
The top view in Figure 47 shows the bolt from above, with the tapering end for insertion into the fingers of the trigger (see Figure 68). The bottom side view gives the diameter at two places and the extra height at the rear end with the important indented nock to be gripped by the bowstring.

in anaerobic conditions. Expertly conserved, it has allowed fine detail to survive, like the two narrow bands 'of some unknown, slightly shiny material...on both sides of the tapering end of the shaft' (Figure 48). These are most likely to be glue for binding feather flights onto the shaft. Sinew was widely used in the Near East for binding flights. It would make sense to regard these two bands as marking the front and rear of the standard three feather flights. The only other possibility is that this

Figure 48 – Bolt 211 (top): narrow glue or paint bands at the rear end. Bolt 416 (bottom) with incomplete shaft.
Bolt 416 has an iron head with an unusually long four-sided point. The total preserved length is 37.3cm, the diameter of the shaft at the front is 1.8cm. It is for a larger size of bolt-shooter than a three-span, and when complete may have been a two-cubit bolt. See Figure 27 and accompanying text.

is paint: Egyptian and other arrows of the period sometimes have rings or dots of paint. These details are mentioned in the 6th Season Yale report p. 453-4. Only the flight feathers are missing. The wood is beech (*fagus sylvatica*), a dense hardwood that resists splintering. The 10.5cm long iron socketed bodkin head is a Manning Type 1. The total shaft length of 50.8cm is close to the standard length of 46cm for a two-span bolt-shooter. In view of the comparatively short range across the river I think it could just as likely have been launched from a slightly smaller Xanten-Wardt machine, a one-and-nine-tenths-span catapult normally shooting a 40.5cm bolt.

Like the group of wooden-flight *manuballista* bolts found at Dura-Europos (Figure 63), it was clearly made with a drawknife or similar trimming tool to achieve the extra height of the shaft at the back end. This extra height at the back end has allowed an indented nock to be created for the bowstring, a feature unconfirmed until this discovery. Most modern bolts are made from lengths of dowel. A second bolt, number 416, with only part of its Spanish maple shaft (*acer campestre*) surviving, looks to be suitable for a larger catapult (see the description below Figure 48). I am grateful to Erik Graafstal for the information and photographs of these finds.

Figure 49 - Two scorpion reconstructions
(left) The Vitruvian three-span *scorpio maior* built by Len Morgan. The battle-shield has been omitted in order to show the rope springs and centre-stanchion. (photograph: the author)
(right) The enemy's view of the Vitruvian *scorpio maior* built by Tom Feeley, incorporating a battle-shield based on the layout of the 1887 Cremona find (photograph: Tom Feeley).

The Qasr Ibrim foreshaft and iron bodkin heads

A distinctive design of bolthead plus shaft has been identified, contemporary with Vitruvius himself. For a few years from 25/4 BC Roman legionaries occupied the fortified stronghold of Qasr Ibrim, poised on high cliffs above the Nile near Abu Simbel (Chapter 10). The finds sealed in the layers of this exciting multi-period site are in amazing condition. They include very large numbers of *ballista* balls. From bolt-shooters there are several iron bodkin points with tangs, with one surviving specially turned and shaped hardwood foreshaft onto which they were mounted (Figure 50). Similar iron heads and incomplete hardwood foreshafts found at Haltern and Vindonissa in Germany have fostered arguments as to whether they were short but complete missiles, perhaps intended for use with crossbows, or whether they were foreshafts whose 'vanes' were spliced onto long mainshafts with flights and shot from catapults.. The author agrees with the latter interpretation by James and Taylor (James and Taylor 1997) based on the Qasr Ibrim example. The mainshafts may have been of softwood, and the hardwood foreshaft would have guaranteed deeper penetration of the iron point before the shaft snapped, a particularly useful characteristic for disabling enemy horses. The expansion in the centre of the foreshaft and the way the 'vanes' start from a sharply cut shoulder are essential to providing a stopping point for the front end of the mainshaft, resisting its forward thrust on impact. The mainshaft end of the foreshaft receives three sawcuts, creating grooves into which the three 'vanes' of the foreshaft can be pushed (see Figure 50 diagram), similar to the present-day method of attaching the cruciform plastic flights to darts. The reported 18.5 mm diameter of the Qasr Ibrim foreshaft is very close to the 18 mm we have calculated for the three-span *scorpio* from the evidence of the aperture in the Cremona shield (Figure 45).

Figure 124 shows Qasr Ibrim heads with one bent on impact. Similar iron points with tapering tangs are now being reported at other sites, such as Alesia and Vindolanda, although these were probably slotted straight into a standard wood shaft. I have identified (Figure 27 right) an object from the Roman fort inside Hod Hill native hillfort, published as a carpenter's bit, as a larger version of this tanged type of bolthead suited to a two-cubit bolt-shooter with a bolt 92 cm long. A similar two-cubit size tanged head has been found in Dumfriesshire (Andrew Nicholson pers. comm.).

Figure 50 - Qasr Ibrim foreshaft and bolthead
(left) Reconstruction of the assembly of a tanged bolthead and hardwood foreshaft onto a softwood mainshaft. (After James and Taylor 1997, Figure 6) **(right)** Foreshaft and bolthead from Qasr Ibrim. Lathe marks are visible on the shaft (British Museum. photograph: the author). See also other tanged boltheads from the site in Figure 110.

Penetration of different types of body armour

Figure 51 - Bolt damage to lorica segmentata
Damage to a replica *lorica segmentata* by a three-span *scorpio maior* bolt at a range of 40m. **(left)** The bolt has smashed through the upper breastplates, forcing them apart and **(right)** punched a square hole (right) in the backplate and distorted it. Of course, there was no *subarmale* (inner jacket) or body inside to absorb some of the impact. (photographed by the author during filming for the BBC)

Figure 52 - Bolt strike on chainmail and helmet
Penetration tests at a range of 40 metres supervised by David Sim. **(left)** Three-span (69 cm) bolts have easily sliced through chain mail, travelling through two heavy leather sheets and foam bodyfiller, and penetrating 18mm plywood. **(centre and right)** David Sim's replica helmet bowl seriously distorted by the first shot of the day. (photographs: the author).

The rope springs and the bolt-shooter's performance

The power of the catapult depended entirely on the quality of the rope springs. According to the artillery writers, animal sinew rope was much preferred as the best. The only clear information on the performance of a bolt-shooter includes the weight of sinew used. A contemporary of Vitruvius, Athenaeus Mechanicus, describes the ranges recorded by the engineer Agesistratos (Marsden 1969, 88): '...*he so outstripped his predecessors that anyone reporting the facts on his behalf is not easily believed. For his three-*

span catapult shot three and a half stades with twelve minae of sinew.' (647 metres with 5.24 kg of sinew rope). Athenaeus implies that to shoot a three-span bolt, 69 cm long and probably weighing about 180 gm, this far was an exceptional achievement, one which beat normally achieved ranges. Agesistratos was a Greek specialist artillery engineer who no doubt tuned his catapult to perfection and chose the time and weather. He may, of course, have used a flight arrow.

The key references are:

> '... *it has been found that the more exercised sinews of an animal happen to have more* **eutonia,** *for example the sinews from the feet of deer [i.e. Achilles tendon], from the neck of a bull ......The spring-cord on the arms is also made of women's hair, because being fine, long and fed with plenty of oil it acquires much* **eutonia** *when plaited, with the result that it does not fall short of the power obtained from sinews.'*
> Heron, Belopoiika, 110 - 112

eutonia is best translated as '*efficiency when stretched*'.

> '*onagers, ballistas and other catapults are of no use unless strung with sinew ropes. However, horsehair from tails and manes is claimed to be useable for ballistas. There is no doubt that women's hair is just as excellent in catapults, as Roman crises have proved.'*
> Vegetius IV. 9

As to the use of women's hair, the number of occasions when women were shorn in the cause of their city's defence has probably been exaggerated. During the siege of Carthage in 147-6 BC Strabo records that '*slave-girls provided hair for the catapults*'. There seems no doubt that a spring made of a skein of sinew-rope gave a markedly superior performance to one of horsehair. The properties of individual or small groups of sinew fibres have been analysed by John Landels (Landels 1978, 107-11), who concluded that the catapult engineers had chosen '*what was probably the best material available to them, since its energy-storing capacity is, believe it or not, higher per unit weight than spring steel.*' Materials scientist Caroline Baillie and master bowyer Steve Ralphs tested short lengths of a variety of ropes for the 2002 BBC *ballista* programme. They confirmed that sinew rope gives back an exceptionally high percentage of the energy stored in it. They also found that horsehair rope was comparatively poor in this respect, a fact which confirms the author's experience with it. Schramm and others tried without success to make rope from sinew. In 1996 it was successfully manufactured for a small catapult by Digby Stevenson, using the backstrap sinew that runs from the loin to the backbone of cattle (Harpham and Stevenson 1997, 13-17). There are now several online videos of preparing backstrap sinew for bow making.

The labour-intensive process involves cleaning and drying the sinew, then pounding it with a mallet, painstakingly splitting it into fine fibres; after soaking these to remove crimps and make them flexible, they are spun into yarn thick enough to be plaited into rope. Heron mentions (*Bel.* 110, above) the Achilles tendon of deer, and Steve Ralphs made a short length of sinew-rope from Achilles tendons for the BBC giant *ballista* TV programme. He found that during the process of drying, the length of the sinew rope shrank by about 10%.

Vindolanda Tablet II, 343 is a letter, possibly dating from 122 AD, from an Octavius who appears to be either a civilian contractor or a centurion responsible for supplying the army with hides, grain etc. (Birley A. 2002, 114-6). The first item on his list is 100 Roman pounds of sinew (32.75 kg). Such a large amount of presumably raw sinew could have been intended for catapult springs, and should be compared to the weight of sinew, 5.24 kg, in Agesistratos' three-span scorpion's remarkable shoot, the only figure we have for the amount of sinew in a catapult. Of course, Agesistratos, being a master artillery engineer, probably succeeded in cramming in more sinew rope than the average three-span had. A catapult use for Octavius' sinew would fit in with the known presence at that date of legionaries at Vindolanda, possibly in preparation for the building of Hadrian's Wall. Sinew ropes gradually lost their efficiency (Philon *Bel.* 61; Heron *Bel.* 110), and would have to be replaced regularly if the catapult was not

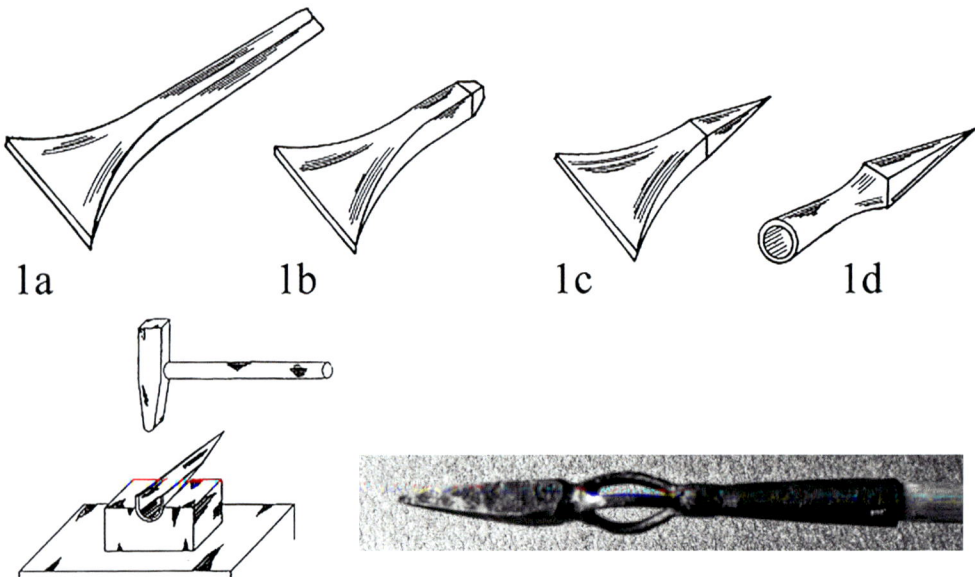

Figure 53 - Artillery boltheads
David Sim's reconstruction of the methods of manufacturing artillery boltheads. 1a to d: Stages in making a standard four-sided bodkin point. **(bottom left)** Creating the socket from Stage 1c. (After Sim.) Bodkin points of a slightly different profile were also standard for mediaeval crossbows. **(bottom right)** Fire bolt for a scorpion, with a cage for inserting flammable material. Replica by Tom Feeley.

to lose power. The evidence that the Xanten-Wardt 0.42 litre size of spring may well have been retained in the new arch strut *cheiroballistra* design may mean that this size of rope spring was known to produce a very effective battlefield performance from an economical amount of the vital sinew-rope.

The quality control of the vast amounts of sinew required may have been achieved under legionary workshop supervision of the expanded workforce type described in Berlin Papyrus 6765 (Bishop and Coulston 2006, 42), using auxiliaries, soldiers' slaves, and civilians; the latter could have included garrison wives and daughters, whose spinning skills would have been ideally suited to processing backstrap sinew or Achilles tendons into rope.

Fitting, tensioning and twisting the rope springs

The Greek engineers had discovered that the energy storing properties of sinew-rope could be best achieved by pre-stretching it, and a special stretching frame with winches is described by Vitruvius (X, x, 12,1) and Heron (*Bel.* 107-10). The skeins of spring-cord were wound around top and bottom iron washer-bars which are slotted into the revolving bronze cylinders traditionally translated as 'washers' (Figures 34 and 40). The end of the rope was tied to one washer-bar and threaded through to the opposite bar. This first strand is stretched, says Heron, until it loses one third of its diameter, then a wooden clip is applied to the rope to retain the tension; successive strands are applied in the same way. Vitruvius says that the tensioned cord is plucked to sound a musical note and that all strands are tuned to the same note. Presumably a tone-deaf engineer would have found it difficult to string a catapult.

When there appears to be no room for any more rope, Heron describes how to get extra strands through the washers by driving blunt iron bars into the skein with a mallet to open up space. As the arms were winched back ready for the release, the rope strings were twisted; the combined tensioning and torsion stored a massive amount of energy in the sinew. The rope springs' capacity to store energy could be maintained by applying a large spanner and twisting the washer-bars and washers towards the front of the catapult. '*If in the course of frequent winding back*,' says Heron (*Bel.* 110), '*the spring slackens, you will twist the washers, as mentioned above* [*Bel.* 101]*, with the iron bar that has the ring*'; a modern reconstruction of such a ring spanner is visible in Figure 43. The description 'torsion catapult' derives from the twisting of the rope springs which was created every time the arms were wound back to the shooting position. It has been suggested that sinew-rope was so efficient that it gave back 90% or more of the energy stored in it by tensioning and twisting.

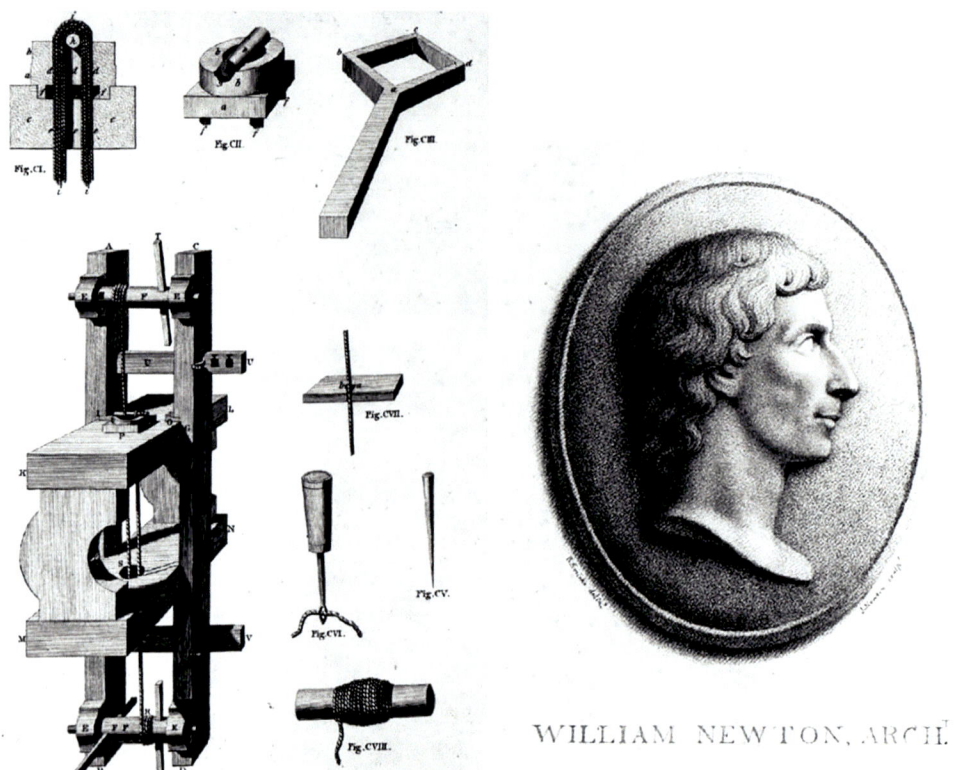

Figure 54 - William Newton's version of the rope tensioning frame described by Vitruvius and Heron

Newton suggested replacements for all Vitruvius' missing diagrams. This, the only surviving portrait, was engraved by his brother from a drawing by Sir Robert Smirke.

Chapter 6

The new design: the metal frame arch strut *cheiroballistra/manuballista*

The Xanten-Wardt spring-frame provides an important link and relationship to the completely new design of bolt-shooter whose first known appearance is on Trajan's Column in Rome recording the campaign against the Dacians of AD 101-102. It would have been impossible to reconstruct this machine from the evidence of the reliefs on the Column, but fortunately a document describing a small version of this design has survived. There have always been hopes of finding 'lost' manuscripts of Greek and Latin authors. Napoleon III's Bibliothèque Impériale employed a manuscript

Figure 55 - Catapult scene on Trajan's Column, Rome (cast in the Victoria and Albert Museum, London)

Two of the five scenes of catapults on Trajan's Column in Rome (Scene LXVI, Casts 164-6), depict events in Trajan's first campaign against the Dacians in AD 101-2. Two machines with the characteristic semi-circular arch strut of the *Cheiroballistra* text **(top left)** are sited on the rampart of a Roman fort. In the **right foreground** a similar machine manned by two legionaries is mounted on a raised platform of logs (photograph: the author).

hunter, Minoïde Minas. On his death a manuscript which included the artillery texts was found amongst his papers, with a note that it had been discovered in 1843 in the library of the Vatopedi monastery on Mount Athos. The information from this and other manuscripts was incorporated into a very fine edition of the Greek artillery writers by the Alsace scholar Carle Wescher, published 'by order of the Emperor' in 1867. Wescher's analysis showed that the Minas manuscript contained the earliest and best copy of the Greek artillery writers. He used it to produce an improved version of the highly intriguing Greek document entitled 'Heron's Construction and Dimensions of the *Cheiroballistra*'; the Latin equivalent of *cheiroballistra* is *manuballista*, 'hand catapult'. It is similar to a modern workshop manual. The document had been first published in 1693 from a single manuscript copy, now known as P, in Louis XIV's Paris Library.

The five pages of the manuscript contain descriptions and coloured illustrations of eight mechanical parts with names beginning with the letter kappa. The names differed from those for the parts of catapults in Heron and Philon's accounts. When first translated the parts seemed to be: Two boards, Locking piece, Field-parts. Pittaria (?), Arch, Ladder, Bronze cylinders and bars, and Two Cone-shaped pieces. Köchly and Rüstow admitted that they could not make sense of the text. One of many suggestions was that more than one machine was being described.

In 1862 the French Professor of Mathematics A. J. H. Vincent proposed that it was a machine of the type described by Philon, powered by compressed air cylinders. His engineer pupil Victor Prou dismissed this as impossible, but in the same year published in collaboration with Vincent a translation of the Greek text interpreted as a catapult powered by metal springs, as is also described by Philon (Prou with Vincent 1862). Vincent showed Prou that the manuscript illustration of the Arch appeared to be the same as the arch strut on the machines on Trajan's Column in Rome. In 1877 Prou published a lengthy article, with accompanying full-size reconstruction of the machine by Albert Piat, as a steel-spring powered bolt-shooter (Prou 1877). He decided that the title 'hand ballista' meant that it had hands holding the springs. So, he made the cast bronze hands seen in the top diagram of Figure 56. No arms could be seen on the Trajan's Column catapults. Prou argued that because small catapults were often called 'scorpions', the arms must have been similar to a scorpion's forward-facing pincers. So, he researched all the different types of scorpions and featured them (Figure 56). This idea that an ancient catapult could have had forward facing and inward swinging arms is a theory which has attracted some support today.

In 1906 the German classicist Professor Rudolf Schneider dismissed Prou's reconstruction as a work of fantasy (Schneider 1906), and boldly, not to say recklessly,

The new design: the metal frame arch strut cheiroballistra/manuballista

declared that because all the parts began with Kappa the text was just a short extract from a technical lexicon listing unconnected parts alphabetically. Curiously, he conceded that some parts might be from catapults, but claimed that the title 'Construction and Dimensions of the *Cheiroballistra*' had been erroneously added by a Byzantine scribe. Schramm accepted this strange, damning verdict and rejected the text, in spite of Tittel's argument (Tittel 1913, col. 1041) that the eight mechanical parts were '... *obviously from the construction of a catapult. It... looks more like one homogeneous description, which is however cut short.*' As a result there was a staggering 94 years following Prou's reconstruction when the *Cheiroballistra* text was ignored.

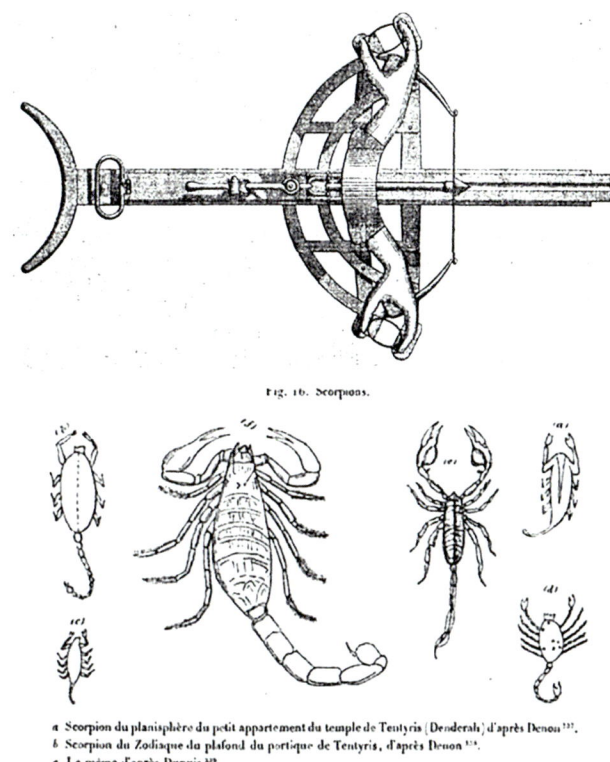

Figure 56 - Victor Prou's cheiroballistra (from Prou 1877 Figures 47 and 16)

Figure 57 - Trajan's Column catapults
(left) Detail of the right-hand rampart catapult in Figure 55. Note the object on the right behind the legionary's helmet, which may show the crescent-shaped fitting (Figure 67) to the rear of the axle of a winch. **(right)** Scene LXVI, Cast 169. Two Dacians on a stockade with a Roman catapult (the top of the semi-circular arch has disappeared). (Casts in the Museo della Civiltà Romana, E.U.R., Rome. Photographs: the author.)

The two Dacians with a Roman catapult in Figure 57 must have captured it either during Trajan's campaign or some 15 years earlier when the Dacians had killed general Cornelius Fuscus and captured the catapults and engineers of *Legio V Alaudae*. Trajan himself recaptured them during his campaign of 101-102 (Dio lxviii 9, 3 and 5; Wilkins 1995, 8). If these Dacians are in possession of one of Fuscus' machines then the introduction of the arch strut design can be brought forward to *c.* 86-7.

Eric Marsden's identification of the machine

In 1971 Eric Marsden reinstated the document, identifying it as a description of a bolt-shooting catapult of a new design, equipped with metal instead of wooden spring-frames, and, as Vincent had pointed out, with an arch strut identical to that on the catapults shown on Trajan's Column. Thus, the machine was established by Marsden as a bolt-shooter issued to the Roman army from the time of Trajan onwards. He completed his reconstruction from the evidence of the text and ms illustrations alone, long before Dietwulf Baatz' identification and publication of the parts discovered in the Danube forts. He chose the higher of the two ms readings for the spring-frames. At my last meeting with him before his tragic early death, he showed me the photographs of the parts which Baatz had sent him, and said that his reconstruction was not correct: 'so we shall have to work out a fresh interpretation'. This he did not live to do, and I have tried to keep up the momentum of his work on this and the other Roman catapults.

Figure 58 - Eric Marsden's cheiroballistra
Eric Marsden with his reconstruction of the *cheiroballistra*, against the background of an arch strut catapult on Trajan's Column. (photograph of Eric provided by Mrs Margaret Marsden, photograph of the Column: the author).

The new design: the metal frame arch strut cheiroballistra/manuballista

Figure 59 - Two carroballistae *on Trajan's Column*
Trajan's Column, Scene XLI. Two *carroballistae* (cart-mounted ballistae) in action, shooting over the heads of advancing troops. (photograph of the Column: the author.)

Marsden's interpretation had been vindicated when in 1974 Gudea and Baatz published several iron catapult parts found in the late Roman forts on the Danube at Gornea and Orşova (Romania). They identified them from the manuscript illustrations as a huge version of the *cheiroballistra*'s arch strut and several sizes of the metal spring-frames (Figures 74 and 75).

Tragically, Eric Marsden died before he could suggest a revised interpretation of the machine based on the new evidence of actual parts. In 1994 I published (JRMES 6, 1995) a reconstruction based on a revised Greek text and translation and the Rumanian finds, and following my construction of a full-size but only semi-working machine in 1991 for display in Tullie House Museum, Carlisle (Figure 60). From this Len Morgan, my professional engineer collaborator and distinguished reconstructor of Roman artefacts, built the fully-operational catapult seen in the following pages. It will soon find a permanent home at Comlongon Castle near Dumfries, where all our catapult reconstructions will be exhibited and will regularly be displayed shooting against targets.

Today there is complete agreement amongst scholars that the *Cheiroballistra* document is indeed a description of a bolt-shooting catapult powered by torsion springs, the washers and washer-bars for the rope-springs being clearly described and illustrated. However, the text breaks off suddenly at the end in the middle of describing the

Figure 60 - The author's first interpretation of the Cheiroballistra *manuscript, built in 1991*
In Figure 60 **(left)** I am testing the catapult, constructed with the help of Grant Hollis of Hollis Engineering, Annan, who manufactured the metal parts. I made the Locking Rings (Figures 81 and 82) of oak painted bronze. Hence the catapult was not capable of taking the strain of full wind-back. **(right)** The catapult mounted for display in the Hadrian's Wall turret in Tullie House Museum, Carlisle, complete with sound effects. It remained there from 1991 to 2023. It will be put on display elsewhere in the Museum, complete with the original design of stand.

arms, and there are other signs that our copies have gaps, such as a sentence with no main verb. Because it does not include a mention of a winch or stand there are some scholars who do not believe that the catapult was a standard winched torsion machine. (It is worth noting that there is no mention of the essential washers in the surviving copies of Vitruvius' description of the scorpion bolt-shooter.) They have attempted to prove that it was a much less powerful one operated as a stomach-bow. Aitor Iriarte (2000, 47-75) proposed to downgrade it further to one retaining the mss figure of Aγ (alpha gamma), 25.6 mm, for the spring diameter, spanned by hand and firing small darts weighing 25 gm, less than a third of the weight of the standard war arrow. It would have been pointless to equip the Roman army with a machine that only offered a feeble performance inferior to that of hand-bows and *arcuballistae* hunting bows (see Appendix A. The Roman origin of the mediaeval crossbow). Such a light dart would not be of much use in a non-military context for bringing down game. '*The object of catapult construction*' Heron and Philon remind us, '*is to project the missile over a great distance and strike a hard blow at a given target.*' Philon advises, '*We must direct most of our research, as we have frequently emphasised, to achieving long range and to hunting down the features of engines that lead to power*'.

Curiously, Iriarte does believe (Iriarte 2000, 68 and Figure 19) that the Gornea frames are from <u>winched</u> arch strut catapults, even though they are much lower in height than the *cheiroballistra*'s: Gornea 2 is 14.4 cm high as against the *cheiroballistra*'s 20.2 cm (Figure 69).

The new design: the metal frame arch strut cheiroballistra/manuballista

There are two irrefutable reasons for believing the torsion *cheiroballistra* to be a standard winched catapult. First, one critical measurement, the cross-section of the *cheiroballistra*'s wooden case, has been increased by about 47% over that of the comparable size of Vitruvian *scorpio*, the Xanten-Wardt catapult, implying that the *cheiroballistra* was generating more, not less power than the previous design of wood-frame winched catapult. The figures for this calculation are: Xanten-Wardt case: 60mm high by 44mm wide (= 26.4 square cm), *cheiroballistra* case: 58mm high by 67mm wide (= 38.8 square cm). This unarguable fact is sufficient to rule out any reconstruction which downgrades its power.

Second, Heron states categorically (*Bel.* 84-5) that torsion catapults developed so much power that the stomach-bow's stomach-bar (Figures 8 and 10) had to be replaced with a winch, with a pulley-system added for the larger machines. On the author's reconstruction the *cheiroballistra*'s draw pull has been measured by Len Morgan at 335 kg, four times that required to draw the most powerful hand-bow. Of course, Iriarte's version, with a small 25.6 mm diameter and 0.13 litre volume of spring (see comparable statistics in the following paragraph), has reduced the power output to such an extent that he can draw it back without a winch.

The transition from *scorpio* to *manuballista*

The total height of the later arch strut *manuballista/cheiroballistra*'s metal spring-frame plus the two washers, 202+32+32=266mm, is identical to that of the Xanten-Wardt *scorpio*, an extremely striking and significant fact that cannot be accidental (Wilkins 1995, Figure 8 and Figure 10a). If we then apply Marsden's emendation of the *cheiroballistra*'s spring-diameter, Bγ (beta gamma) 2 1/3 dactyls = 45mm instead of the mss reading Aγ (alpha gamma) 1 1/3 dactyls= 25.6mm, the rope spring diameters of both machines are also identical, and so are their spring volumes at 0.42 litres. This matches Vegetius' observation (IV. 22) *scorpiones dicebant quas nunc manuballistas vocant* ('what they used to call scorpions they now call *manuballistae*'). It implies that the designers of the arch strut *manuballista* had the Xanten-Wardt size of *scorpio minor* in front of them as a starting point, and retained the Xanten-Wardt size of springs in their final version. On this evidence the Xanten-Wardt *scorpio minor* may be regarded as the predecessor and a progenitor of the metal frame arch strut *manuballista/cheiroballistra*. The reason may well be that this size of rope spring was known to produce a very effective battlefield performance from an economical amount of the vital sinew-rope. For example, the amount of rope required for ten arch strut bolt-shooters of the larger Lyon size (Baatz and Feugère 1981, Figure 2) with springs of 1.86 litres would be sufficient to arm 44 *manuballistae/cheiroballistrai* with 0.42 litre springs. A general faced with a choice between these two might well be tempted to choose the latter.

All this fatally undermines Iriarte's version. In any case Iriarte's version is unauthentic because he has knowingly overridden the ms evidence by inserting the Ladder's tenons through the Pi-brackets (see Figure 71 and comments in the opening Section 'Archaeological discoveries and recent research').

The *cheiroballistra's* all-metal spring frame: appraisal of this radically different design

Eric Marsden was right to list the *cheiroballistra/manuballista* as a new category of bolt-shooter, and not as the Mark VI version of the Hellenistic machine, because the *cheiroballistra* and its other arch strut brothers do not simply represent a replacement of wood by metal, but a new design of frame, arms and stand, as seen on Trajan's Column. The 'safety' of the centuries old, well-tried Hellenistic plated wood version was abandoned, and the challenge undertaken to create entirely in metal a spring frame whose characteristics produced increased power and vastly improved handling in battle. The compressive and torsional stresses to which the new material would be subjected would have been unpredictable and potentially dangerous. A bold, thorough programme of strain testing during the development stage is implied.

The author of the *Cheiroballistra* text is Greek, and the expertise of Greek artillery engineers, like that of Greek architects and doctors, continued to be consulted. However, the design of the completely new frame, arms and stand must surely have been initiated by Roman artillerymen voicing their criticisms of the traditional wood frame's deficiencies, and suggesting a thorough reappraisal of user requirements and improved weapon characteristics, all based on their many years of experiencing the harsh realities of the battlefield in foreign and civil wars. Their criticisms would have included the difficulty of close-combat aiming. Key demands are likely to have been more power to increase the range, a clearer view of the enemy, and faster rewind to increase the rate of launch. The arch strut bolt-shooter's increase in arm arc of 15° to 20° produced a notable increase in power; and its superb unimpeded 'widescreen' field of view, plus the semi-circular arch aiming aid, gave an enormous boost to the artillerymen's accuracy of shot and so their confidence in the heat of battle, as they faced hordes of rapidly approaching enemy. Faster rewind would have been facilitated by the double winch handle technique of the Cupid Gem (Figure 85), and speeded even more by the crank handle-handspike of the Elenovo find (Figure 87).

The success of this family of bolt-shooters must be credited to the demands of the legionary *ballistarii*, and to the superb quality of Roman metalwork produced by the legionary *fabricae* and, in the later Empire, by the specialist workshops of Autun (central France), Trier (west Germany) and other centres appointed by the Roman high command to manufacture *ballistae*. The legionaries now had a weapon offering:

The new design: the metal frame arch strut cheiroballistra/manuballista

Increase in power

This was achieved by allowing the arms to be pulled back through a greater arc and thus store more energy in the rope springs. Two changes made this possible: first, the two spring-frames are spaced three times further apart than on the Vitruvian bolt-shooter, creating a much-increased arm arc. Second, by borrowing the *palintone* style of frame used on traditional stone-throwing catapults: the Vitruvian bolt-shooter had an all-in-one *euthytone* wooden frame with the stanchions in a straight line, limiting the arc through which the arms could travel before they hit the stanchions (Figure 40). To produce the greater power needed to project stones (Chapter 7), two separate spring-frames were used, with their stanchions staggered out of line in *palintone* configuration, in order to increase arm travel.

The idea of modifying a *palintone* stone-thrower to shoot a bolt was not new: Caesar (Civil War II 2, 2) recounts how in 49 BC the citizens of Marseille used very large ones to shoot enormous spike-tipped beams 12 Roman feet (3.56 metres) long at his siege lines; he says that they passed right through four thicknesses of wickerwork screens before sticking in the ground. Agesistratos used a *palintone* stone-thrower to hurl a four-cubit (1.84 metre) bolt four stades (740 metres) (Marsden 1969, 205).

Improved field of view

For the first time there was a superb, unimpeded field of view, plus an artificial aid to precision aiming. The Vitruvian wooden frames blocked off a clear line of sight along the missile to the enemy, and prevented direct aiming along the missile and slider, as Heron advises. especially when the enemy were at close quarters. It was then difficult to judge the elevation, because the initial flight of the missile has a fairly flat trajectory and it becomes lost behind the frame. It was necessary to aim over the top of the frame or along the side, particularly for sizes larger than the Xanten-Wardt (Figure 61), and the operator must estimate how far to raise or lower the case to achieve the desired elevation.

The *cheiroballistra* has no centre-stanchion, affording a totally unimpeded view of the enemy and the ability to aim along the shaft of the bolt, which is far more accurate and has always been the traditional technique of archers throughout history. Heron advises (*Bel*. 86, 7-8), '*We shall shoot at the target by looking down the length of the case*'. The open framework allowed the approaching enemy to be followed at all times however much he swerved, and in close combat the final shot could safely be delayed until a hit was certain and the bolt penetrated with maximum force. We find that the open frame also allows the missile to be followed throughout its trajectory, from the moment when it leaves the case, to about 200m distance, so that corrections of elevation and direction can be more easily estimated.

Figure 61 - Aiming the three-span scorpio and the cheiroballistra *(photographs: the author and Margery Wilkins)*
Figure 61 **(left top)** is the artilleryman's view of the Vitruvian three-span *scorpio maior*, with the all-in-one spring-frame and battle-shield blocking his view. **(left below)** The same view of the *cheiroballistra*. There is no centre stanchion to block the view and the two spring-frames are three times further apart. **(right)** During filming for Adam Hart-Davis' programme 'What the Romans Did for Us' Len Morgan is aiming the three-span *scorpio* at the BBC target on the site of Hod Hill's Iron Age Hut 37, at a range of 150m. 48 of the 50 bolts he shot during filming landed within a 10m circle, the diameter of Hut 37.

Our experience with reconstructions of the machine supports Marsden's important suggestion (1971, 227-8) that accurate medium and close-range aiming could have been assisted by framing the target in the semi-circular arch on the upper strut. The arch is the most characteristic feature of the new design, and the very fact that this upper metal beam has been physically weakened by the insertion of the arch emphasises the priority given to improving ease and accuracy of aim. This is the first known sighting aid on a weapon in the western world; there is no evidence for the existence of backsights and foresights before the 18th century.

In one of the first outings of our reconstruction, at Corbridge Roman site in 1997, a novice *ballistarius* produced a 'Robin Hood' result: his first bolt pierced the plywood target at a range of 60m, his third shot gouged a groove along the shaft of the first.

Faster speed of rewind and consequent increased rate of missile launch

If the twin handle and ratchet pullback design shown on the Cupid Gem (Figure 59) was introduced with, or at a later stage for the arch strut design, it would have provided a faster rewind and the ability to launch more bolts in a given time. Equally, when the crank handle-handspike was introduced for catapult use, possibly, as Kayumov and Minchev suggest (2010, 337 and Figure 15), at the end of the 2nd century, this would allow an even faster initial wind back. Tom Feeley and Len Morgan's brilliant

The new design: the metal frame arch strut cheiroballistra/manuballista

reconstruction of the crank handle-handspikes on a *carroballista* size of arch strut machine (Figure 88 below) has enabled us to compare full rewind time with the push-pull double ratchet system shown on the Cupid Gem on an identical reconstruction by Len Morgan. The timed results are that the crank handle system is one third faster than the push-pull system, which is itself some 20-25% faster than the basic winch operated by separate handspikes.

When faced with the life-threatening situation of a rapidly approaching enemy, as envisaged in Arrian's battle plan against the Alans (Chapter 11), the artilleryman would definitely not appreciate the extended rewind time of an 'inswinger' with an extra 30° or so to winch back using the weaker leverage of much-shortened arms (Chapter 14 and page 190).

Elimination of ricocheting bolts

A further advantage of the new design was the elimination of the risk of bolts missing the centre stanchion's aperture, striking the centre-stanchion, and ricocheting back at the crew (a not infrequent hazard today). Fortunately, the bolt in Figure 62 did not ricochet.

Figure 62 - Misfire (still frame from video: the author)

A three-span bolt on a Vitruvian replica in Figure 62 has missed the narrow aperture in the centre-stanchion and stuck in the wood of the stanchion (note that the bowstring is fully released). This type of misfire cannot happen on an arch strut bolt-shooter (seen here in the foreground) because there is no centre-stanchion.

The new design of bolt, with aerodynamic and horrific wounding properties

The aperture for the bolt in the old centre-stanchion had to be small to avoid weakening the stanchion. The omission of this stanchion removed this restriction on the width of the bolts that could be launched.

The bolts shot by these arch strut catapults were almost certainly of the improved design with wooden flights found at Dura Europos on the Euphrates, and described by the historian Procopius (*Goth*. i. 21 in Marsden 1971, 246-7). The open framework allowed a standard iron bodkin head to be mounted onto a stockier, expanding shaft with wooden instead of feather flights (Figure 63). One of the advantages of this evil-looking missile over the older dowel-shafted bolts is that its shaft has wounding properties of its own: its expanding profile will create a constantly enlarging entry wound, ripping muscles and crippling, if not killing, men and horses. If it penetrates up to the flights, being of wood these will open the wound further and act like barbs, making removal of the missile extremely difficult and painful. This was first confirmed by Wild Dreams Films'

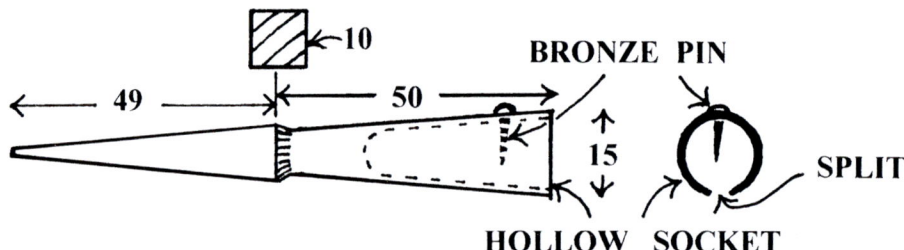

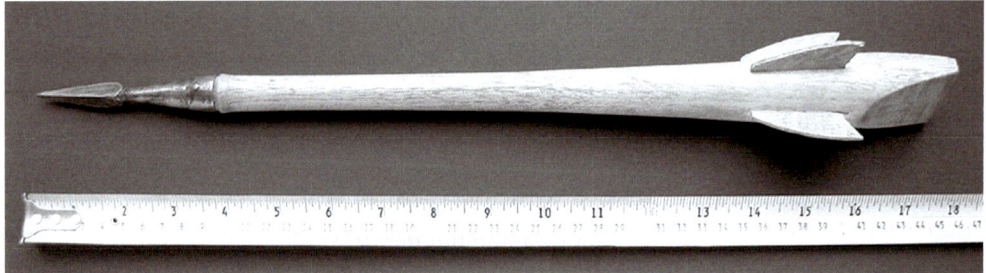

Figure 63 - Dura Europos bolt replica
A replica by the author of one of the bolts suited to the *cheiroballistra* design of catapult, found at Dura Europos in Iraq (Baur and Rostovtzeff 1931, 72-3 and Plate IX). The bodkin head was made by John Anstee. The **top diagram** is of the bolthead of the one surviving complete shaft, redrawn from the excavator's field card in the Yale University archive, with measurements in millimetres.

slow-motion film of a *cheiroballistra* bolt penetrating a large melon (Figure 65): the bolt caused lateral splits in the melon and left a gaping exit 'wound'.

More detailed and dramatic proof of its horrific wounding properties came from our filming for Blink Films in 2020, when we were asked to shoot at a large block of ballistic gel, as used for replicating the effect of bullets on human flesh. The result confirmed the expanding entry wound caused by the profile of the shaft, but unexpectedly revealed a massive split, shaped like a bow wave from a boat, opening up on both sides of the shaft. One can imagine the terrible effect of this bolt as it ripped into men and horses.

This design of bolt is usually straight and stable in flight, because the Concorde-like profile of its shaft is stabilised by the slipstream of passing air. This is confirmed by American wind tunnel tests (Foley, Palmer and Soedel 1985, 84-85) which have also proved its low ratio of drag to mass, and the fact that its tail, tapered to fit between the claws of the trigger (Figure 68 below), reduces the turbulence at the rear. Our Vitruvian three-span bolts made of straight 18 mm dowelling lack this stability, sometimes rocking in flight.

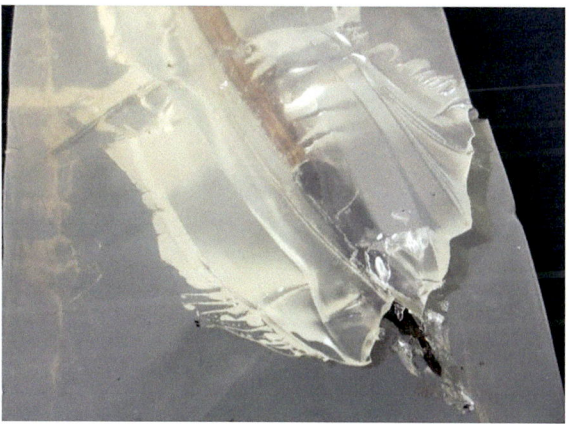

Figure 64 - The horrific strike effect of this bolt on ballistic gel simulating human or animal flesh Len's marksmanship ensured a first time hit, using one of Tom Feeley's replica Dura Europos bolts. The lateral splits on both sides of the shaft are shaped like the bow wave of a boat. (photographs: Margery Wilkins).

Figure 65 - Slow-motion film of a cheiroballistra *bolt penetrating a large melon*
The bolt also caused lateral splits in the melon and left a gaping exit 'wound' (Still frame from: 'Ancient Warfare' by kind permission of Wild Dreams Films Ltd).

The *Cheiroballistra* manuscripts

The *Cheiroballistra* text is unique among the artillery documents in that it is the only one to give exact sizes for the components, rather than a system of proportions based on the size of the spring-hole. This may have been because it was the archetype, and that larger versions like the *carroballista* or the Lyon or Orşova examples were developed from it. There were 16 dactyls (finger-widths) of 19.3 mm to the Greek foot. There are four main manuscripts of the text: M (the Minas manuscript) and P in the Bibliothèque Nationale in Paris; V in the Vatican Library and F in the Austrian National Library in Vienna. F is a poor copy with an incomplete text and crude versions of the diagrams.

The following is necessarily a very brief commentary on the eight parts described in the manuscript. The name of each part begins with the Greek letter kappa.

Kanones duo (two boards) i.e. Case and Slider

The Vitruvian wooden case and slider were retained; they were not the sort of components that were suitable for translating into metal. The 47% increase in the cross-section of the *cheiroballistra*'s case, over that of the comparable size of Vitruvian *scorpio*, the Xanten-Wardt catapult (calculation on page 75) proves that the *cheiroballistra* was more powerful than this previous design of wood-frame winched

The new design: the metal frame arch strut cheiroballistra/manuballista

Figure 66a - The Cheiroballistra manuscript
The first of the five pages of the Vatican copy with the diagram of the Case and Slider.

Figure 66b - The Cheiroballistra manuscript
Page Two and the diagram of the Arch, Ladder and T-Clamps. (photographs: by kind permission of the Vatican Library).

The new design: the metal frame arch strut cheiroballistra/manuballista

catapult, and invalidates any reconstruction which retains the mss reading of 11/3 dactyl (25.6mm) spring diameter. In addition, the case was slightly lengthened to 22 times the spring diameter, as against the 19 times of the Vitruvian *scorpio*, in order to cope with the increased arc of the arms' travel. This lengthening required and accounted for a small amount of the increase in cross-section.

There is one new component that does not feature on the Vitruvian *scorpio*, the crescent-shaped fitting attached to the end of the case. Some scholars who have tried to downgrade this catapult to a stomach-bow believe that this fitting was mounted horizontally as the equivalent of Heron's stomach-bar. However, the crescent-shaped fitting lacks the end handles of the stomach-bar (Figures 8 and 10), and two labels on M's diagram indicate that the case and fitting are drawn in side view, with the fitting mounted vertically. Scene LXVI, Cast 164 on Trajan's Column (Figure 57 left) may show

Figure 67 - Operating the cheiroballistra
The Roman Military Research Society operating the Mk I *cheiroballistra* reconstruction for Adam Hart-Davis' seminal BBC programme 'What the Romans Did for Us'. Seneca's very first shot for the cameras was a bull's eye. (photograph: the author)

the crescent-shaped fitting to the rear of the axle of a winch. Note the interpretation of the use of the crescent-shaped fitting at the rear of the case in Figure 67. It does not work as a shoulder piece.

Kleisis (trigger mechanism)

The manuscript diagram and Heron's diagram of the *gastraphetes* (Figure 10) is the best evidence for artillery trigger design; no actual trigger parts of any catapult have been identified to date. In Figure 68 the claw (*chele*) imitates the archer's two fingers on the bowstring The snake (*drakontion*) is the release lever that locks the claw over the bowstring. This is one of the nicknames given to the parts, probably by the *ballistarii* who operated the catapults. It does have a snake-like appearance and when pulled back to release the claw it delivers the deadly sting. The tapering rear end of a Dura Europos type of bolt (see also Figure 63) is fitted into the claw. The goalpost-like Pi-bracket (*pittarion*) which straddles the trigger is unique to this catapult, and may have been used to push the slider forward. It does not work as a backsight, as has been suggested. Heron advises aiming along the case. Foresights and backsights do not appear until the 18th century. The author has made the bowstring of flax, following the design of bowstring described for mediaeval crossbows (Payne-Gallwey 1903, 110 – 113).

Figure 68 - Cheiroballistra *trigger mechanism*
Len Morgan's realisation of the author's interpretation of the trigger mechanism, based on the text instructions and ms diagram (photograph: the author).

The new design: the metal frame arch strut cheiroballistra/manuballista

Kambestria (Field-frames)

This is the simplest of solutions for a metal replacement of the Vitruvian spring-frame: two rings to hold the washers, spaced apart by two bars, one of which has the semi-circular arm recess (Figures 69 and 70 below). As Marsden noted (1971, 222), the Greek name is the equivalent of the Latin *campestria*, frames for use in the field (*campus*), a reminder that these catapults were primarily intended for use on the battlefield rather than for static defence. The word *pittarion* used to describe the brackets attached to the bars is not found in any modern dictionary, and only when the Gornea and Orşova frames were identified was it realised that it simply means 'a bracket shaped like the Greek letter pi (Π)'. They look like miniature goalposts. For the author's suggested use for the Pi-brackets see Figures 81 and 83. The Gornea No. 2 field-frame in Figure 69 is only about two thirds the height of the *cheiroballistra* frames; its ring is slightly narrower but the internal diameter of 3 dactyls is the same as that calculated for the *cheiroballistra* (below page 90). The manuscripts' reading of a mere 1 1/3 dactyls (25.6 mm) for the internal diameter of the washers and the diameter of the *cheiroballistra*'s rope-springs would make the catapult 'little more than a toy' (Marsden). The majority of scholars have always supported Marsden's corrected reading of 2 1/3 dactyls (45 mm) for the rope spring diameter, exactly the same as that now known for the earlier wood frame Xanten-Wardt scorpio minor, which may now be regarded as inspiring the choice of rope-springs for the *cheiroballistra*.

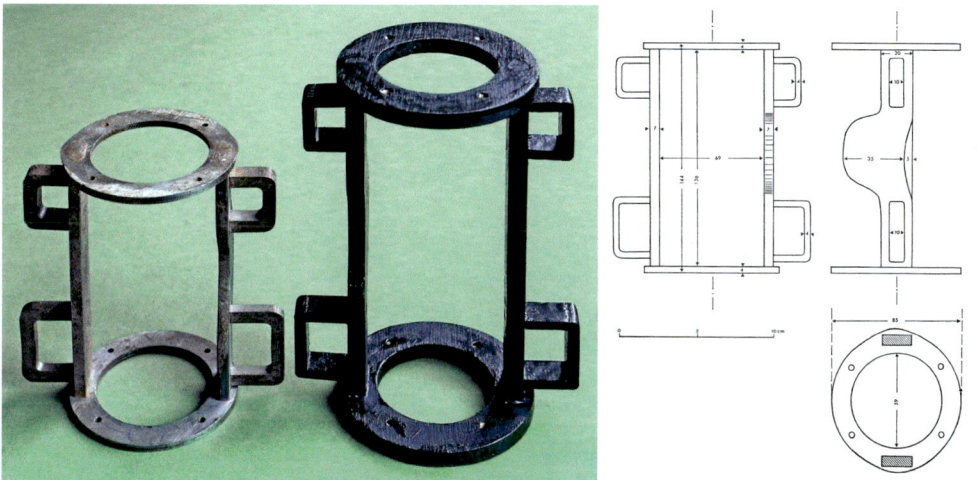

Figure 69 - Field-frame replicas and Gornea example
(left) Replica of the Gornea No. 2 field-frame (left), by Len Morgan, and that of the *cheiroballistra* (right) by the author, showing the pi-brackets. **(right)** Drawing of the No. 2 iron frame. It was discovered in the tower of the fort at Gornea on the Danube, dated to the late 4th century AD (After Gudea and Baatz).

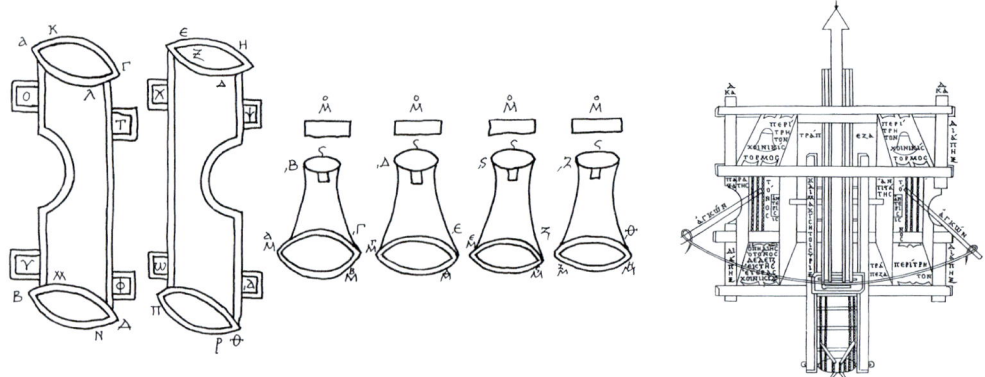

Figure 70 - Codex M's Field-frames and Heron's ballista diagram
(left) Codex M's drawing of the pair of Field-frames, the Washers and the Washer bars. **(right)** Heron's diagram of the *ballista*, which we have proved (Chapter 14) to have standard outward swinging arms. (diagrams after Wescher)

As on the wood frame bolt-shooters, there is a semi-circular cut-out for the arms, which the *Cheiroballistra* manuscript diagram (Figure 70) of a pair of field-frames portrays as being on the outer sides of the frame assembly. On the author's reconstruction they therefore retain this, the same position as on the wood frames. It is important to note that the cut-outs are drawn in exactly the same position and manner in Heron's diagram of the *ballista* (Figures 70 and 102) where they are definitely on the outer sides of the frames.

Pittaria (Pi-brackets)

The Greek word *pittarion* is unique to this manuscript. The surviving Field-frames all confirm that it must mean 'a fitting in the shape of the letter Π [Pi]'. In Codices M, P and V some of the Pittaria have been coloured in, emphasising that they are indeed shaped like the Greek letter. On the Gornea Frames the *pittaria* seem to have been fitted by inserting their tenon ends through mortise holes in the Bars and peening their ends as if they were rivets. All these Field-frames have smaller *pittaria* at what is presumably the top. Gornea No. 3 appears to have its straight Bar assembled upside down, a mistake which could be taken to indicate that an unskilled apprentice was involved in its construction or repair. (Iriarte suggests the mistake was made during the 20th century conservation process.) The larger Field-frames from Orşova and Lyons also have smaller *pittaria* at the top. The possible reason for the difference in size is discussed below. The surviving text does not mention this difference. The outer section of the Bulgarian Elenovo Pi-Brackets is only 2mm thick (Kayumov and Minchev 2010, 336 and Figure 13), and they are, like those of Gornea No. 2, riveted to the side-stanchions, not welded.

The new design: the metal frame arch strut cheiroballistra/manuballista

Figure 71 - Junction of the Ladder tenons and the Field-frames (models and photographs: the author)

I have followed the evidence from Gornea in reconstructing the *pittaria*. There is a lack of regularity and precision in their construction: if the top or bottom *pittaria* are viewed as pairs, they do not always match up accurately in size or alignment with one another: Gornea No. 2 illustrates this (Figure 69 right), as do the Orşova and Lyon Frames (Figure 72). I take this lack of alignment to imply that their function did not require them to be shaped or positioned with absolute accuracy, hence my suggestion that they are 'wedge-frames' (Figure 81).

Iriarte (2000, 62) is in agreement with this identification: '*Wilkins makes a bulls-eye when he identifies the strangely non-matching pitaria...as wedge-frames.*' However, his use of the wedges is to fix the Arch and Ladder tenons through the Pi-Brackets (2000, Figure 13), even though he says that, '*As Wilkins remarks, it is impossible to pass the tenons of the 'ladder', which are 3d<actyls> apart, through the pitaria in the kambestria.*' (2000, 58). And again, '*I must admit that Wilkins is right when he says that the 3d<actyl> apart bars of the 'ladder' are not in the best position to pass their tenons through the lower pitaria of the field-frames. This is the weakest point in my reconstruction.*' (2000, 62).

This is admirably honest of him, but, as pointed out in the section 'Archaeological discoveries and recent research', by ignoring the problem and forging ahead and inserting his Arch and Ladder tenons through the Pi-Brackets, against the clear mss evidence, he has misled and inspired a large army of followers who have produced 'tenons through Pi-Brackets' reconstructions, and who concomitantly do not believe that there are missing parts. In fact, I am aware of only one other reconstruction which believes that there are missing parts. The latest reconstruction, by Kayumov and Minchev of the Elenovo field-frame, using CAD illustrations, follows this 'tenons through Pi-Brackets' error, and adds inward swinging arms.

The ms measurements result in the position in the above Figure 71, with the tenons resting on and level with the Rings of the Field-frames. The Figure 71 left wooden mock-up of the Ladder shows that its tenons are not spaced sufficiently apart (ms measurement 3 dactyls = 58mm) to insert into the Field-frame's Pi-brackets. The ms says that the Field-frame stanchions are spaced 3½ dactyls (67mm) apart, so the Ladder tenons would have had to be spaced 67mm + the thickness of the two stanchions apart. This amounts to at least 67+9+9 = 85mm. The ms figure of three dactyls for the spacing of the Ladder tenons is the same as the inner diameter calculated for the Rings of the Field-frames. This results in the tenons resting on and level with the Field-frame Rings.

The Field-frame Rings

The diagram in the codices (Figure 70) and the actual Field-frames from Gornea (Figure 69), clarify the meaning of the *Cheiroballistra* text, and leave only the Ring dimensions to be debated. A large iron Field-frame complete with iron Washers and Bars was found at Lyon, France (Figure 72). Its spring diameter of 74mm (three Roman inches) is the same as that of the old Vitruvian three-span *scorpio maior*, but its height is less. The spring volume is 1.86 litres compared with the Vitruvian 2.24 litres, but the 15° to 20° increased arc of the arms may have made up all, or nearly all of the difference. The Lyon catapult would have been of a size suitable for mounting on a cart as a *carroballista* (Figure 59). The photograph from Baatz and Feugère 1981 (Figure 72 centre) shows that to achieve the *palintone* configuration, the Lyon metal stanchions are in an offset position on the rings, with extra metal added to the outer edge of the rings in order to achieve this. This offsetting would be impossible to achieve on the much smaller area of metal on the rings of the Gornea Field-frames and of the *cheiroballistra*; on the latter I have suggested that the offset was probably achieved by means of the differing lengths of the beams of the Ladder recorded in the mss (Figure 78). Three other large Field-frames have been found, one at the late 4th century AD fort at Orşova on the Danube, estimated spring diameter 69-70 mm (Figure 72 centre top), a bronze version in Morocco, estimated spring diameter 80 mm, and an iron version from Elenovo, Bulgaria, estimated spring diameter 54-5mm (Figures 73 and 74).

All the codices give B (beta=two) dactyls for the breadth (euros) of the Field-frames' Rings. Marsden takes this as their internal diameter and argues convincingly that since the Bars are three and a half dactyls apart, this figure is suspect (for his full reasoning see Marsden 1971, 223-4). He suggests ΓC (3½) dactyls (1971, 224 - 5). Baatz demonstrates (Gudea and Baatz, 1974, 63) that this figure for the internal diameter should be further reduced to three dactyls, because that is the actual measurement of the Gornea Rings, and also because Marsden was not to know that the mortise holes for the Bars are cut half way across the Rings.

The new design: the metal frame arch strut cheiroballistra/manuballista

Baatz must be right, and all I would change is that by regarding this ms measurement of breadth as the <u>outer</u> diameter and changing the ms reading from B to E (two to five) dactyls, then we arrive at exactly the same result through a palaeographically easy correction. With this one correction, these measurements for the Rings, 5 dactyls outer diameter and 1 dactyl Ring width, and distance between the Field-frame Bars 3½ dactyls, produce proportionally similar measurements to Gornea No. 2 (Figure 69), for which the outer Ring diameter is 4.4 dactyls (increasing to 4.7 dactyls next

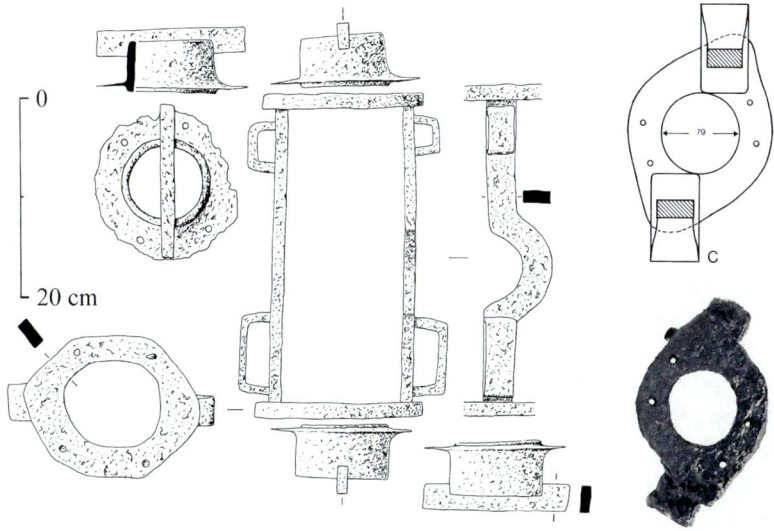

Figure 72 – Lyon and Orşova Field-frames

Figure 72 shows **(left)** the large iron Field-frame complete with iron Washers and Bars found at Lyon (diagram after Baatz and Feugère). **(right bottom)** The photograph from Baatz and Feugère 1981 shows that to achieve the palintone configuration, the Lyon metal stanchions are in an offset position on the rings. The same offset arrangement **(right top)** on the Orşova frame (after Baatz).

Figure 73 - The Elenovo Kambestrion (photographs: Kayumov and Minchev)

The Elenovo group of metal objects was found in 1962 in the vicinity of the village of Elenovo in Southern Bulgaria. It included the crank handle-handspike mentioned in the Introduction (Figure 2 and fully illustrated in Figures 87 and 88) and a probable catapult arm (Figure 80).

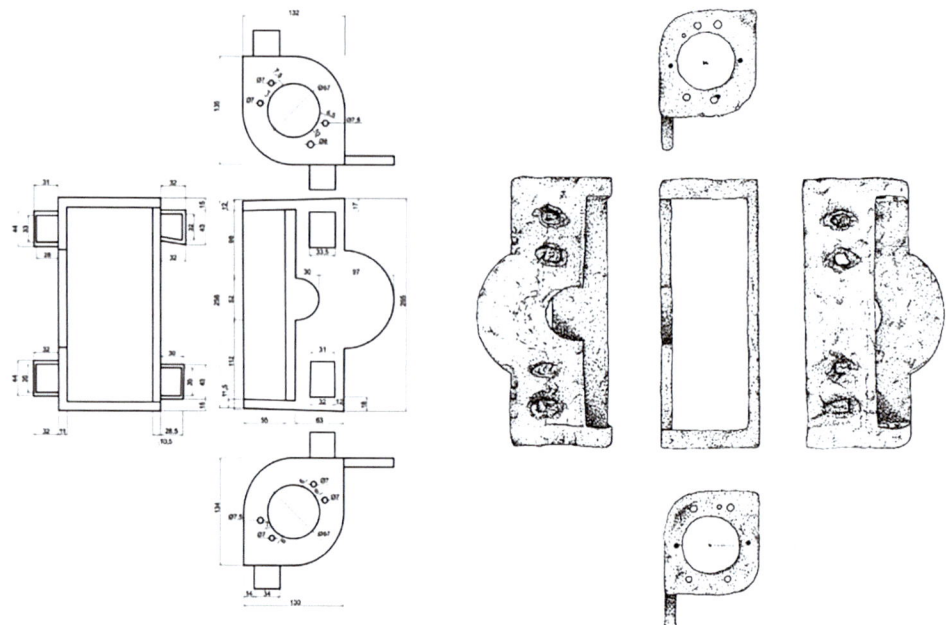

Figure 74 - Kambestria *from Elenovo and Sala (Morocco)*
(left) Diagram of the Elenovo *kambestrion* (Kayumov and Minchev). **(right)** The *kambestrion* from Sala in Morocco. Both have offset stanchions which allow a palintone configuration of the catapult frame. (diagram from Dietwulf Baatz).

to the mortise holes for the Field-frame Bars), the Ring width is 0.88 dactyl, and the distance between the Field-frame Bars 3.58 dactyls. Out of the total of 53 numerals in the *Cheiroballistra* ms this figure for the measurement of Ring breadth and that for the internal diameter of the Cylinders/Washers are two of the only three which appear to require amendment, - a tribute to the remarkable accuracy with which this text has been transmitted. If only the text of Vitruvius had survived as accurately! Of course, we can see why Iriarte was tempted to produce an explanation allowing <u>all</u> the ms measurements to be correct, including the 11/3 dactyl (25.6mm) spring diameter. No Greek or Latin manuscript has survived in copies with 100% accuracy.

Kamarion (Arch)

This component is named after the semicircular arch which is easily recognised in the mss diagram, and which features on all the catapults on Trajan's Column. Figure 76 compares Codex M's drawing, the profile of the actual arch found at Orşova (Figure 75) and the author's version drawn from the detailed measurements given in the Greek text. They are virtually identical: the ms drawing has survived surprisingly undistorted and the Orşova Arch emerges as a 2.75 times enlargement of the *cheiroballistra*'s, but proportionally identical to the mss drawing. It presumably belonged to a very large tower-mounted static version of the machine, a Mons Meg of its day, with a spring diameter of about 6½ dactyls, five Roman inches. Note also the two holes to left and right of the

The new design: the metal frame arch strut Cheiroballistra/Manuballista

Figure 75 - Orşova iron Arch strut (photograph: Dietwulf Baatz)

Arch. These may be explained as for holding a cord stretched across the arch with a movable bead to adjust the aim, as found on medieval crossbows. It has been suggested that they may be rivet-holes for holding vertical struts to strengthen the framework. If this was so, it would rule out inward-swinging arms.

The broken Orşova tenons display two extremely important details: first, a rectangular hole and the start of a second, which are not mentioned in the text. The author interprets them as holes to receive rectangular lugs which would somehow lock the Arch in position. Similar holes are assumed to have been present on the tenons of the Ladder (Figure 78). Second, the ends of the Orşova Arch's tenons are reduced to almost wafer thickness (Figures 75 and 76), which can only mean that they were sandwiched

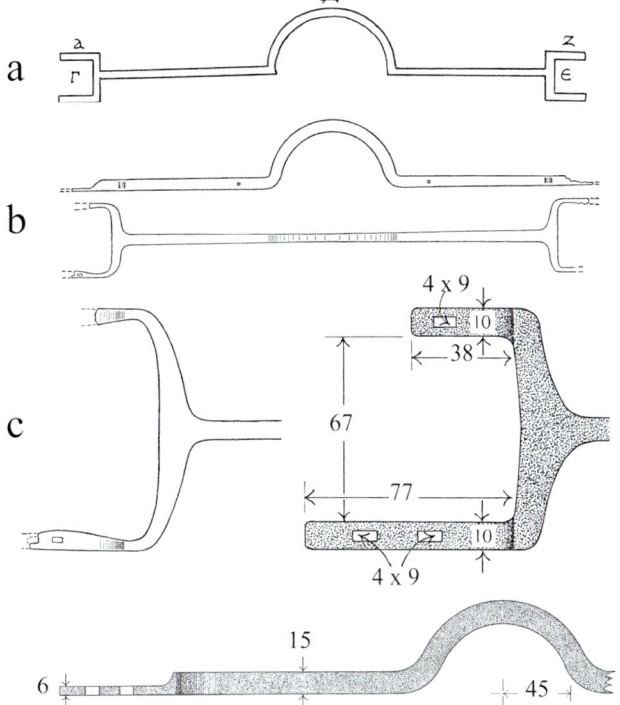

Figure 76 - The Cheiroballistra Arch

Figure 76 shows the Codex M version. Either the forked tenon ends, or the semi-circular Arch, have been turned through 90°, there being no tradition at that time of using two drawings to show views from two different planes. 76b: Drawings by Baatz of the huge Orşova Arch (after Gudea and Baatz 1974). 76c: Detail of Baatz's drawing of the broken forked end of the Orşova arch showing the surviving rectangular hole and start of a second. The author's stippled drawings are of the reconstructed forked end and a side view of the Arch, with measurements in millimetres.

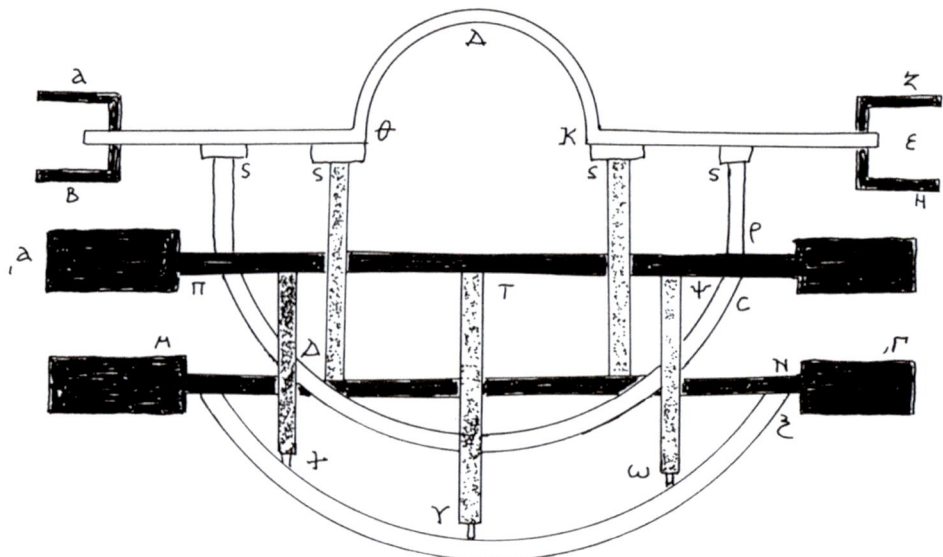

Figure 77 - Codex P Arch and Ladder diagram
Figure 77 is of Codex P's combination drawing of the Arch and Ladder (see also V's version in colour on Figure 66b above). The side beams and tenons of the Ladder are coloured red in the manuscript, black in this tracing. Centuries of copying have resulted in several errors: for example, the small T-clamps that lock the Case to the Ladder have become confused with the rungs of the Ladder and joined to the Arch (diagram from Wescher).

between other components. Most reconstructions simply thicken them up and insert them and those of the Ladder into the Pi-brackets, thus ignoring the manuscript measurements which make such insertion impossible, as demonstrated above (Figure 71 left). The lug holes are not shown in the mss, but are based on the evidence of those on the Orşova arch (Figure 76). There is a gap in the mss description of the thickness and length of the tenons, but they are probably the same as the Arch tenons, the longer one being 4 dactyls (77mm) long, the shorter 2 dactyls (38mm). I am very grateful to Aitor Iriarte for a suggestion leading to a better reading for the text at this gap.

Klimakion (Ladder)

To date no Ladder has been found. Therefore, it has been reconstructed from the text measurements plus the limited help afforded by the diagram (Figures 77 and 78). Whereas the Arch is a comparatively slender top brace between the two field-frames, weakened mechanically by the inserted arch, the Ladder is an extremely strong component that takes the weight of the Field-frames, spring-cord, Washers and Arms, locks them rigidly in position, and resists the enormous force of winding back and the shock of recoil, the violence of which is only revealed by slow-motion photography (2005 TV programme: Wild Dream Films' *Ancient Discoveries* 'Ancient Warfare').

The new design: the metal frame arch strut cheiroballistra/manuballista

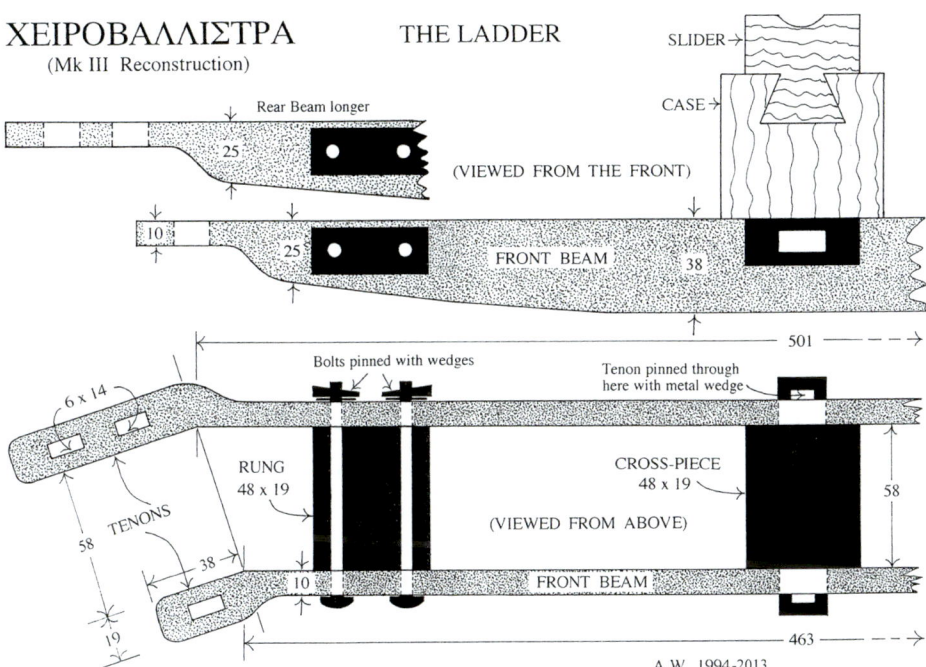

Figure 78 - Reconstruction of the main frame. The Ladder
The Figure 78 photograph is the view from below of the Morgan-Wilkins reconstruction, showing the Field-frames turned towards the front by the angle of the Ladder's tenons, allowing the arms extra forward movement. **(bottom)** The text description and measurements, translated from dactyls to millimetres, are converted into my diagram; the longer rear beam creates the 18° forward turn of the tenons.

The two beams forming the side rungs of the Ladder are of different lengths. My reconstruction allocates the longer beam as the rear one; its additional length allows the tenons to turn at an angle of 18° to the beams (Figure 78): this will set up the staggered or offset palintone position of the Field-frame bars, affording the arms vital extra degrees of travel and extra torsion of the rope-springs. To achieve the palintone configuration on larger Field-frames such as those discovered at Lyon, Orşova and Elenovo, the metal stanchions are in an offset position on the rings, with extra metal added to the outer edge of the rings in order to achieve this (Figure 73). The small Gornea and *Cheiroballistra* field-frame rings do not have enough metal to accommodate offset stanchions.

Kulindroi chalkoi (bronze Cylinders/Washers) and *Kanonia* (Bars)

There does not seem to be any major difference between these and the earlier design of Washers. The ms reading for their internal diameter, the spring diameter, is a puny 11/3 dactyls (Aγ, alpha gamma, 25.6 mm). This has become the source for a major controversy. Aitor Iriarte's reconstruction is based on retaining this rope-spring diameter. It has always been unconvincing, because that very small volume of rope (0.13 litres) could never produce an effective amount of ballistic energy, whatever its use, military or civilian, that Iriarte believes it to have been. Marsden's verdict was that a sinew bundle of only 11/3 dactyls diameter would mean that '*the cheiroballistra would be little more than a toy*'.

Marsden's second point was, '*that the springs would be far too small in relation to the numerous other components for which Heron provides certain, or reasonably certain, measurements.*' Support for this verdict comes from a comparison of the ratios of spring diameter to height. This is 1:5.9 for the Xanten-Wardt, 1:6 for the Ampurias and Caminreal scorpions, and an estimated 1.7 for the Cremona machine. Iriarte's is a spindly spring with an uncharacteristic 1:10.9 ratio. Subsequent writers have agreed and support Marsden's palaeographically straightforward correction to 21/3 dactyls (Bγ, beta gamma, 45mm), giving a spring volume of 0.42 litres.

Marsden pointed out that the two iron stanchions corresponding to the Vitruvian wooden side- and centre-stanchions are 3½ dactyls apart, and that Vitruvius (X.10.2) records that the distance from both stanchions to the spring-hole should be only ¼ h. This was essential to prevent the field-frame rings being dished in by the enormous compressive force of the prestretched rope-spring transmitted through the washers. On this basis Marsden argued that the ms reading for the internal washer / spring diameter should be corrected to 21/3 dactyls. The archaeological finds have vindicated Marsden's correction: on the Gornea No. 2 field-frame, for example, the critical ¼ h. distance is indeed maintained (Figure 69), as it is on the larger frame from Lyon (Figure 72).

The new design: the metal frame arch strut cheiroballistra/manuballista

Final vindication of Marsden's amendment now comes from the very striking fact that the Xanten-Wardt springs have exactly the same Washer to Washer height as the *cheiroballistra*, and the same spring volume when reconstructed with Marsden's brilliant amended reading of 21/3 dactyls. It implies that the designers of the arch strut version had the Xanten-Wardt size of *scorpio minor* in front of them as a starting point, and retained the Xanten-Wardt size of springs in their final version.

It could be that this size of rope spring was known to produce a very effective battlefield performance from an economical amount of the vital sinew-rope. The amount of rope required for 10 bolt-shooters of the Lyon size with springs of 1.86 litres would be sufficient to arm 44 *cheiroballistrai* with 0.42 litre springs. A general faced with a choice between these two might well be tempted to choose the latter.

On this evidence the Xanten-Wardt *scorpio minor* can be regarded as the predecessor and a progenitor of the arch strut *cheiroballistra/manuballista*. This lineage is confirmed by Vegetius' observation (IV, 22), '*scorpiones dicebant quas nunc manuballistas vocant*' ('what they used to call scorpions they now call manuballistae'.

Konoeide duo (two cone-shaped parts) i.e. Arms

One of the apparent aims of this new design was to replace as many wooden parts as possible with metal. However, an arm consisting solely of a slim bar of metal would not have had the bulk to be gripped satisfactorily by the spring-cord. The mss text describes and the illustrations (Figure 79) show these bars joined to thick tapering cones which provide the necessary bulk. Unfortunately, our copies of the text stop in mid-sentence, just as they are about to mention the length of the metal bars and possibly the total length of each Arm. The author's study of Vitruvius' stone-thrower (next chapter) concludes that it had the same arm length as the Vitruvian bolt-shooter, i.e. seven times the spring-hole diameter. That this is also the likely length for the *cheiroballistra* Arms was shown by tests on our reconstruction: with the spring framework mounted at the front of the Case, an Arm only six times the spring-hole diameter proved to be far too short for the length of the Case, and one of eight times is too long. To this may be added the evidence of the Bulgarian find (see next paragraph).

There are two versions of the mss drawings, perhaps because at some point the original was lost in some copies. It is important to note that in both versions of the mss drawings the bars appear to stop well short of the inner ends of the cones where they are held by a ring, but project beyond the outer ends. In view of the missing information from the text, I felt obliged to follow these two points in making our first reconstructions of the Arms. When we made the cones in wood they snapped at the inner unsupported end. Therefore, subsequent versions of the cones are in steel (Figure 83). However, I suffered a breakage of a bar at the junction with the cone, which is always going to be a point of weakness. We have decided to follow the evidence of the Elenovo bar (Figure 80) which implies the existence of an alternative design of arm which avoids this weakness and is much lighter. The length of the bar is 37.1cm, for a washer with an estimated internal diameter of 54-5mm; this suggests that the length was 7 times the spring cord diameter and that the bar ran the full length of the cone. We have made an alternative version of the Arm to match this evidence, using a casing of wood and creating a much lighter Arm (see Tom Feeley's reconstruction in Figure 81). Codex V's version seems to show a dovetail groove in the cones for the insertion of the bars, but the text describes the grooves as 'quadrilateral'. This approach to the Arms, with the metal arm encased in wood throughout its length, was followed by Iriarte and Miks in their reconstructions. The bars have been reconstructed by Len and Tom in steel tempered with the resilience that Philon describes in the case of 'Celtic and Spanish swords' (*Bel.* 71). This alternative version of the Arms is much lighter: Len's arm of this type as estimated for the Lyon size frame is 1.40kg as compared to 2.19kg; Tom's is 1.75kg as compared to 2.25kg. There will be much less inertia to be overcome with the lighter arms, and that will guarantee an increased range. Tom's wood encased the metal Arm on all sides (Figure 81), Len on three sides. The Elenovo arm is slightly curved. Is this its original profile? If so, it would have offered the same advantage of an increased angle of pull-back as Vitruvius' curved scorpion arms (Figure 40).

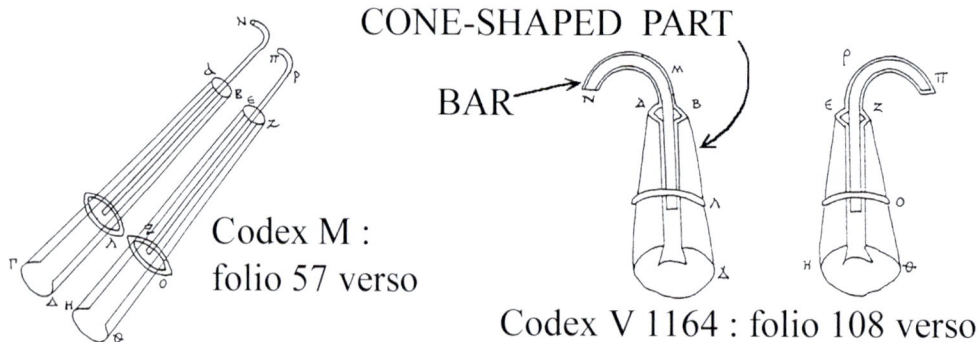

Figure 79 - Codices M and V diagrams of the Arms (diagrams after Wescher)

Figure 80 - The Elenovo arm (photograph: Kayumov and Minchev)

The new design: the metal frame arch strut cheiroballistra/manuballista

Figure 81 - Tom Feeley's version of the alternative wood encased Arms (photograph: Tom Feeley)

Missing parts

Iriarte rightly points out (2000, 60) that the surviving mss text does not explain how the catapult's parts are to be joined up. As demonstrated above (Figure 71), the Ladder tenons cannot be inserted through the Field-frames Pi-brackets. Therefore, disappointingly for modern researchers, there must be missing parts which linked the tenons of the Ladder and the Arch to the Field-frame Rings.

With the pull-back calculated as about one third of a tonne, the torsional forces affecting the *cheiroballistra*'s framework must have required extremely robust linking parts which locked Ladder and Arch to Field-frames absolutely rigidly. Because the Washers rest and revolve on the outer surfaces of the Field-frames rings, both Arch

Figure 82 - The position of the Ladder on the Field-frames. See text below (photograph: the author)

and Ladder tenons must have been somehow clamped to the <u>inside</u> surfaces of the Field-frames rings.

My suggested solution (Figures 82 and 83) is the cast bronze locking 'rings' (krikoι krikoi: remember that all the ms names of parts begin with kappa!) which provide the missing mortise slots for the Ladder and Arch tenons, have projecting lugs which lock into the rectangular slots on the tenons, and wrap tightly round the Field-frame stanchions where they enter the Rings. The author's wooden preliminary model in Figure 82 shows the nearer tenon of the Ladder resting on the inside surface of the bottom Field-frame ring, ready to be clamped between the proposed bronze locking rings and the rings of the Field-frames. The far tenon is already clamped by one half of the locking rings. The author is holding the nearer half locking ring <u>upside down and wrong way round</u> to show the mortise groove for the tenon of the Ladder. Rectangular lugs in these grooves fit into the rectangular holes in the tenons, a system based on the evidence of the holes in the Orşova arch tenons. The two slots at the top of this nearer locking ring fit tightly onto the base of the stanchions. This proposed design for the missing connecting parts ensures that the tremendous twisting force of the catapult's action is applied horizontally to the junction between the stanchions and Field-frame rings. A similar set of locking rings clamps the tenons of the Arch to the underside of the upper Field-frame rings.

Our two-part Arm in Figure 83 has a mild steel 'cone' and a tempered steel bar. It follows the mss drawing which appears to show the bar extending beyond the cone. The author made the bowstring of strands of whipped flax (hemp was not available), following the method used on mediaeval European and early Chinese crossbows, and described by Sir Ralph Payne-Gallwey (1903, 111-13). Heron (*Bel.* 110-11) advises that it should be plaited from the strongest sinews. The author's suggestion for the missing parts is the set of bronze locking rings which butt up against the stanchions. The inner ends of the tenons of the Arch and the Ladder can be seen entering the mortise grooves provided by the locking rings. Pairs of hardwood wedges in the Pi-brackets press down on iron C-rings which hold the two halves of each locking ring together; the locking rings clamp the Arch and Ladder's tenons to the inside of the Field-frames' rings. The 18° turn of the Field-frame created by the Ladder is also visible.

All modern reconstructions of the *cheiroballistra* of which I am aware follow Iriarte and join the Arch and Ladder to the Field-frames by passing their tenons through the Field-frames' Pi-brackets. This is ruled out by the mss measurements: to enable the tenons to be inserted into the Pi-brackets, they would have had to be spaced much further apart. The dramatic loss of thickness of the Orşova arch tenons must also be taken into account (Figure 75). The fact that the mss do not say how the Arch and Ladder are joined to the Field-frames leads to the inevitable, disappointing

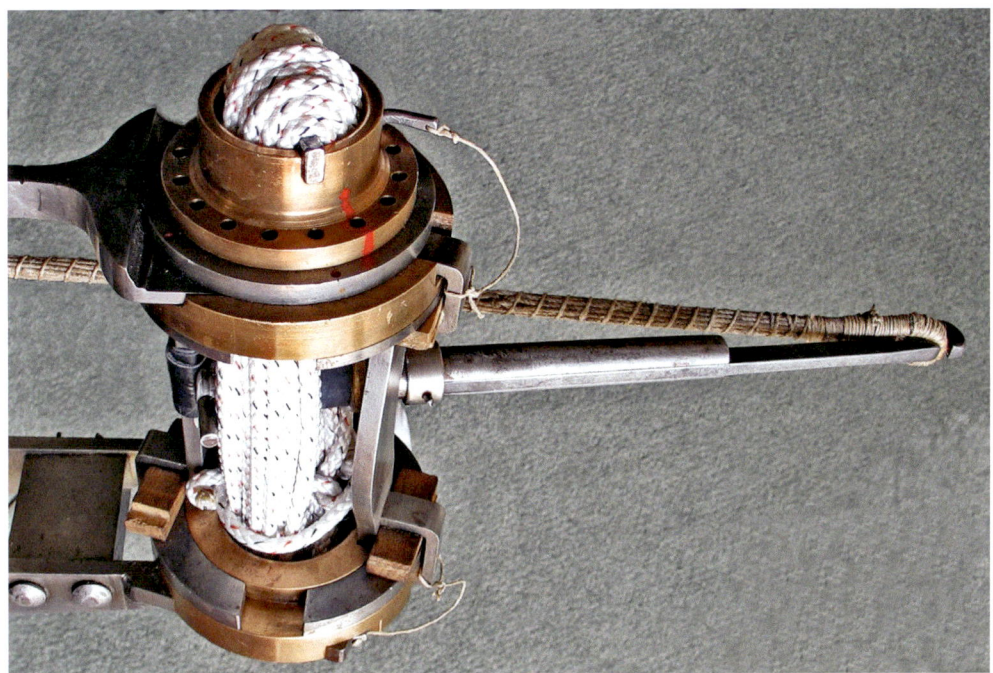

Figure 83 - Left hand end of the reconstructed cheiroballistra
The left-hand section of the reconstruction of the *cheiroballistra*'s frame, seen from the front, with the first interpretation of the Arms (photograph: the author.)

conclusion that there are missing parts, and that both Arch and Ladder tenons were somehow clamped to the inside surfaces of the Field-frames rings.

Even if my proposed answer to the missing parts turns out not to be the system used by the Romans, I hope that the relative positions of the main components that it creates, with the Arch and Ladder tenons clamped to the inside of the Field-frame rings, proves to be correct.

Ease of dismantling and repair

The failed attempt to replace the case on the Xanten-Wardt machine (Figure 37) confirms that if a split developed in this size of the Vitruvian wooden design, or if it was damaged in battle, it could not be repaired on the spot, and it would have to be taken to a workshop and dismantled.

In contrast, the *cheiroballistra*'s metal spring-frame appears to have been built up from a number of bronze or iron components, each of which was linked to the next by removable pins or wedges, or by the interlocking of tenons with mortises. No rivets were needed. It would therefore have been possible to repair a machine with a damaged spring-cord on the battlefield and without the need for an armourer, because

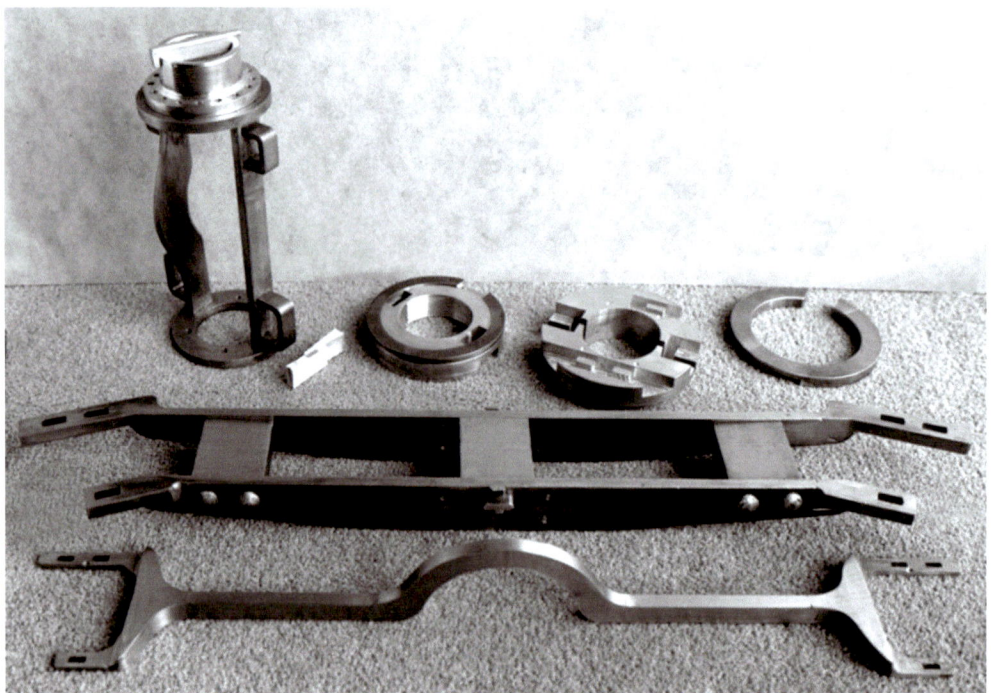

Figure 84 – The author's version of the cheiroballistra *frame parts*
The author's version of the parts of the front frame, realised in metal by Len Morgan (photograph: the author.)

after knocking out the wedges and removing the locking rings, each individual Field-frame could be replaced by a spare ready fitted out with its pretensioned spring-cord, washers and arm. The shorter tenons on Arch and Ladder make it easier to remove and replace the Field-frames.

Winch and stand. Double ratchet handles.

The Cupid Gem is a plaster cast of a carved gem from a finger ring, the only known side view of a Roman catapult. The gem cutter has created a witty version of Cupid firing an arrow, by arming the god with a weapon of enormously greater amorous fire-power than a bow. Cupid is grasping a winch handle in each hand, the advantage of this double ratchet system being a faster speed of rewinding, with a continuous hold kept on both handles, and a push-pull action of the arms keeping the winch axle turning almost continuously. This results in an increased rate of missile release, of particular importance to the artilleryman faced with hundreds of horsemen charging towards him. Was the double ratchet system introduced with the arch strut machines? Clearly cut details include the bowstring, the winch rope, trigger, winch pawl, universal joint, and pinned joints on the stand. This type of stand is not the Philon or Vitruvian tripod type (Figure 46), and seems to be the same as those on

The new design: the metal frame arch strut cheiroballistra/manuballista

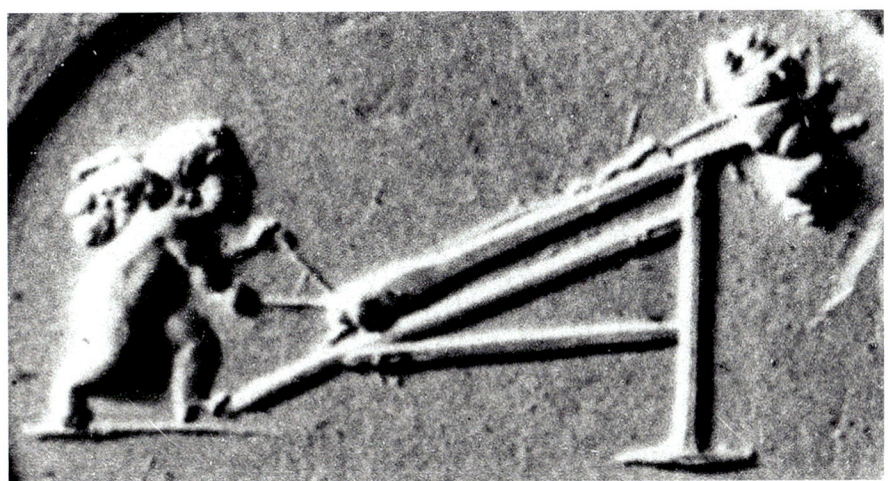

Figure 85 - The Cupid Gem
A plaster cast of a carved gem from a finger ring, the only known side view of a Roman catapult (photograph Dietwulf Baatz).

Figure 86 - Reconstruction of the cheiroballistra
The author's reconstruction of the *cheiroballistra* realised by Len Morgan, with two-handle winch system based on the Cupid Gem (photograph: Margery Wilkins).

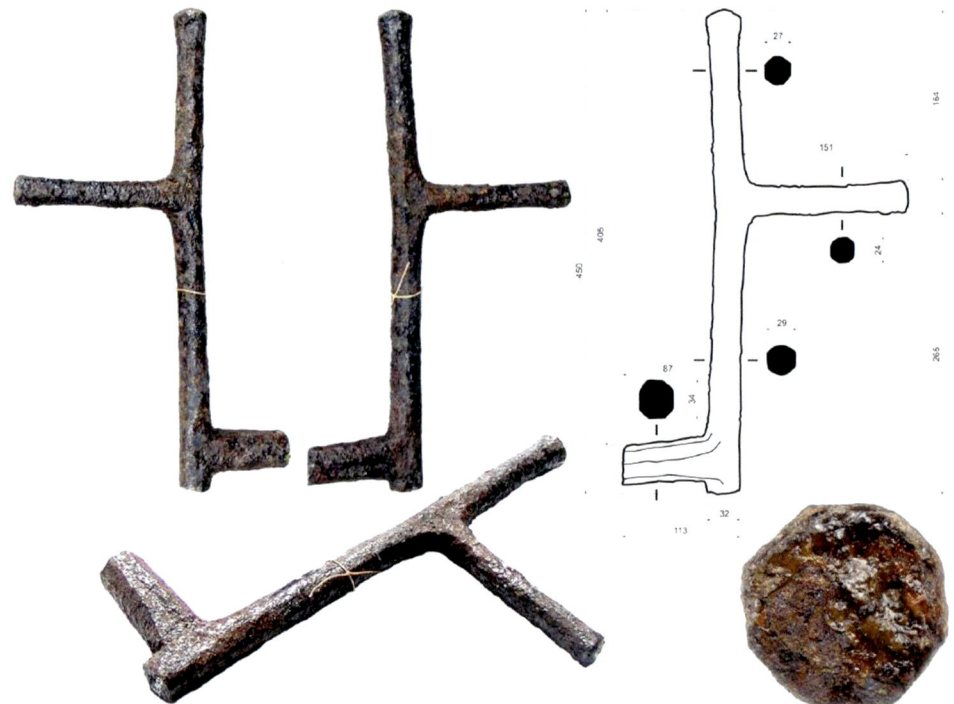

Figure 87 - The Elenovo crank handle and handspike (illustration from Kayumov and Minchev 2010)

Trajan's Column. Therefore, the gem almost certainly shows a bolt-shooter of the arch strut type. The stand is based on the evidence of the Gem and the catapults on Trajan's Column. Straight side ratchets were chosen instead of the circular ratchets of Vitruvius' *scorpio*, partly because this is a safer system if the winch rope breaks, and partly to be able to test and refute the theory that this machine could be spanned as a stomach-bow. A circular ratchet system is just as likely.

The Elenovo combined crank handle and handspike is part of the Elenovo hoard, along with the *kambestrion* ('field frame') from an arch strut catapult and the probable arm. The author agrees with Kayumov and Minchev that this is very possibly from the winch system of this catapult. The fact that it is part handspike reenforces the identification as a catapult part, because only a long handspike can achieve the final few centimetres of pulling the bowstring back to full tension. Len Morgan found that the final 5.5cm of pullback required an extra 27.4kg loading. The existence of crank handles in Roman times has been doubted by many writers, but Kayumov and Minchev record five finds of such handles from sites in the Roman Empire, dating from the mid first century to the middle of the sixth century AD.

The superb reconstruction by Tom Feeley and Len Morgan of the Elenovo crank handle-handspike in Figure 88 has been fitted as a pair to Tom's Lyon size arch strut

THE NEW DESIGN: THE METAL FRAME ARCH STRUT CHEIROBALLISTRA/MANUBALLISTA

Figure 88 - Reconstruction of the Elenovo crank handle-handspike (photographs: Tom Feeley)

catapult, probably the size depicted on the *carroballistae* scene of Trajan's Column (Figure 59). Tom made the important discovery that these handles do not foul the bowstring, even at the end of pull back. Also they result in an even faster wind back than the double ratchet handles system shown on the Cupid Gem (Figure 85). After Tom's death we field tested Tom's recontruction against Len Morgan's with its Cupid Gem type double winch handles. These tests proved that the crank handle-handspike reduced reloading time by one third. The two photographs on the right are of the winch axle which links with the eight-sided shafts of the handles.

Tactical positioning of the *manuballista*

Vegetius lists (II.15) the *manuballista* as a catapult operating behind the front line of the *antiqua legio* ('legion of earlier times'), and sometimes positioned in the fifth line along with *carroballistae* (III.14) thus confirming its power as more than enough to clear these lines. He also mentions *manuballistarii* on a siege tower (IV.22), joining in with slingers, archers, *arcuballistarii* and others to drive defenders off the walls. *Arcuballistarii* were probably artillerymen using *arcuballistae,* non-torsion crossbows, perhaps descended from Greek stomach-bows, or of the same design as the hunting crossbows shown on the two reliefs in the Musée Croatier at Le Puy-en-Velay, France (Figures 161 and 162). These may be the same as the weapons referred to by Arrian's instruction (*Ars Tactica*, 43.1) to train cavalrymen to shoot missiles 'not from a bow, but from a machine'. This could have been spanned by hanging it from the horns of the saddle and using leg power. See Appendix A on the Roman ancestry of mediaeval crossbows.

The protective metal spring-frame covers

There are several incidents during the wars of the early Empire which prove that a wooden catapult was easily damaged or set on fire by enemy action, as in the siege of Jerusalem (Josephus, *Bell. Jud.* V, 279 - 280, 286–7). The old-style battle-shields only covered the front of the oil-soaked (Philon *Bel.* 61) inflammable springs (Figures 37 and 45). All the arch strut catapults on Trajan's Column are depicted with covers over their spring-frames looking rather like the old-fashioned biscuit barrels with domed lids (Figure 89), a detail not found in the text. They would have provided the rope-springs with some protection from damage by fire or axe, and partial protection from the weather, except for the slit left for the traverse of the arms. Tom Feeley has produced impressive reconstructions of these covers (Figures 93 and 94 below), tailor made to fit neatly around the spring-frames.

Fire arrows had been shot from wood frame catapults since Hellenistic times, for example by the Rhodians against Demetrius in 304 BC. See James 1983. Fire-bolts (*malleoli* Figure 53) and fire-spears (*falaricae*) are described by Vegetius as being launched from larger *ballistae* against wooden siege-towers (IV, 18) and ships (V, 44). They could be more safely launched from this new design because of the metal framework and the spring covers.

Figure 89 - Trajan's Column: ballista *on cart (photograph of the Column: the author)*
On Trajan's Column, Scene LXVI a legionary soldier is accompanying either a *carroballista*, or perhaps a standard cart transporting a Lyon size *ballista*. There is no forward-projecting slider on this scene because the catapult is not being used in battle.

The new design: the metal frame arch strut cheiroballistra/manuballista

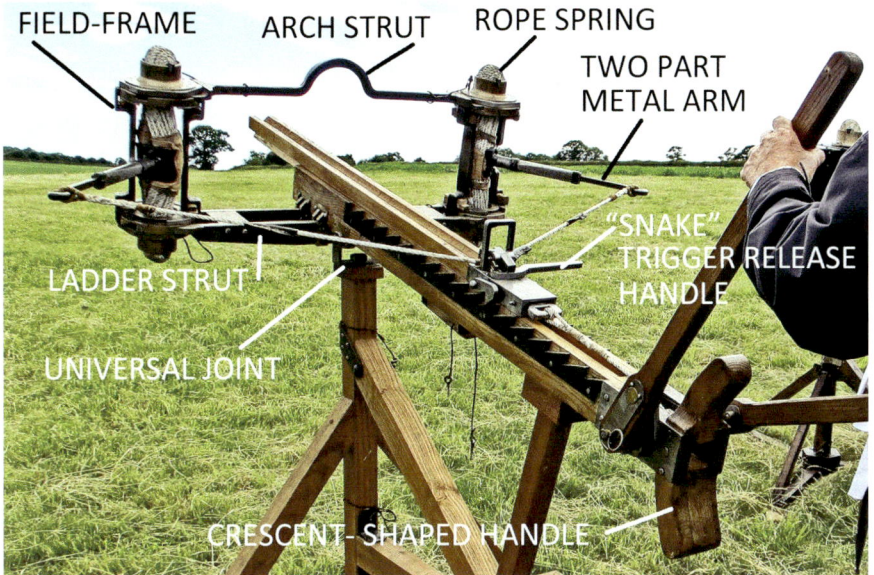

Figure 90 - *The Lyon size of arch strut catapult: Tom Feeley's version*
The *cheiroballistra* design enlarged to the Lyon size with a 74mm spring diameter. Tom later (Figure 81) changed these two-part arms to the Elenovo design (photograph: Margery Wilkins).

Figure 91 - *Adding extra torsion to Len Morgan's Lyon size catapult* (photograph: Margery Wilkins)
Early testing of Len's version fitted with Elenovo type arms. This Lyon size is very heavy and requires four men to lift and assemble it onto its stand.

Because the artistic convention of the Column enlarged the human figures in relation to their surroundings, it is impossible to be certain what size the Column's arch strut catapults are. The two cart-mounted ones, *carroballistae*, (Figure 59) are likely to be larger versions, possibly of the Lyon size, with a spring diameter of 74mm, the same as that of the wood frame three-span *scorpio*. Tom Feeley and Len Morgan have each laboured long and hard to build an arch strut bolt-shooter of this calibre, based on an enlargement of the *manuballista*. Its weight limits its portability, and it requires a team of four to lift it into position on its stand (Figure 91). Undoubtedly this size would also be ideal for mounting on battlements in a defensive role as in Figure 97 top, or setting up in log emplacements (Figure 55), as well on the carts as *carroballistae*.

The arm 'sleeves'

The bas relief of the *scorpio* on Vedennius' tombstone (Figure 35) is so accurate that the sculptor must have had one in front of him. In contrast, there are so many variations in the Column's depictions of the arch strut catapults that the sculptors were obviously unfamiliar with this new design. Sir Ian Richmond's suggestion (Richmond 1936, 3) that they were working from sketches made on campaign could be valid in this case.

The Column's artists were consistent in showing the protective covers on all the catapults. On the two *carroballistae* and the Scene LXVI *ballista* or *carroballista* there is a stubby feature projecting outwards from the covers. (Figures 92 and 96). When Tom Feeley had completed a trial set of the covers, 'cans' for short, he found that the slot cut in them to allow the arms free movement had to be quite long and wide, exposing the rope-springs and arms to possible attack or effects of the weather. I suggested to him that he should

Figure 92 - The Arm 'sleeves'
(left) Close-up of Trajan's Column Scene LXVI, showing one of the protective 'cans' and the outward projecting feature discussed in the text (cast in E.U.R.). **(right)** Tom Feeley's superbly crafted realisation of the 'cans' and the projecting sleeve-like feature (photographs: the author).

The new design: the metal frame arch strut cheiroballistra/manuballista

Figure 93 - Two reconstructions of the cheiroballistra *(photographs: the author)*
(left) The Morgan-Wilkins reconstruction of the *cheiroballistra* ready to shoot a replica of the Dura Europos bolt (Figure 63). **(right)** Tom Feeley's impressive reconstruction, now on display in the Roman Army Museum at Carvoran on Hadrian's Wall.

Figure 94 - Tom Feeley's framework. Impact of bolt on lorica segmentata
Tom Feeley's reconstruction shows the double advantage of the suggested 'arm sleeves' (photograph: the author). **(right)** The impact of a 70gm bolt shot from the Morgan-Wilkins *cheiroballistra* at a range of 50m at replica *lorica segmentata* 1.25mm thick (photograph: copyright Stuart Williams).

try adding the projecting features as possible 'arm sleeves' to reduce this exposure. He confirmed that they performed this function and also produced a quite unexpected major bonus in that they solved a general problem that we have never been able to eliminate – that of the arms jumping upwards at the end of their travel (Figure 94).

Figure 95 - Penetration tests of bolts (photographs: the author)

The left image in Figure 95 is of damage by bolts from the larger Lyon size arch strut catapult at a range of 40m for David Sim's penetration tests. The *cheiroballistra's* Dura Europos design of bolt (centre) has punched through David's hand-forged replica shield boss, penetrating through the 18mm ply behind. It would probably have reached the hand holding the shield. A *cheiroballistra* bolt (right) has forced its way through replica *lorica segmentata*, sliding up the joints in the steel plate and penetrating into the chest of the legionary. A similar result has been achieved against scale armour. The cataphract horse scale armour from Dura Europos, exhibited in the British Museum's *Life in the Roman Army* exhibition, would have offered little or no protection against catapult bolts. (still-frame by kind permission of Wild Dreams Films Ltd).

Did the arch strut design have inward swinging arms? The evidence of Trajan's Column.

The ms diagram of the pair of *cheiroballistra* field-frames shows the semi-circular cutouts for the arms on the <u>outside</u> of the frames (Figure 70). Heron's diagram of the *ballista* (Figure 102), which had outward swinging arms, draws the cutouts in the same manner and position, and clearly shows the arms as swinging to the outside. Heron's states (*Bel.* 100.2) that the arms recoil *outwards* (*eis to ektos*); the Greek for *forwards* would have been *eis to prosthen*.

On three of the arch strut machines on Trajan's Column, the two *carroballistae* and the *ballista* on a cart (Figure 96), there are projections on the <u>outside</u> of the 'cans' which cover the spring-frames. Such additions to the outside would not be required for inward-swinging arms (Chapter 14, Figure 156 for diagrams by Iriarte and Lewis). The clearest depiction of these projections is in Scene LXVI, and has been interpreted by us as 'sleeves' which add protection and also control of the recoil of the arms (Figure 94). There are no signs of arms on the Column's catapults today. Originally, they may have been added in paint or possibly as metal fixtures.

The new design: the metal frame arch strut cheiroballistra/manuballista

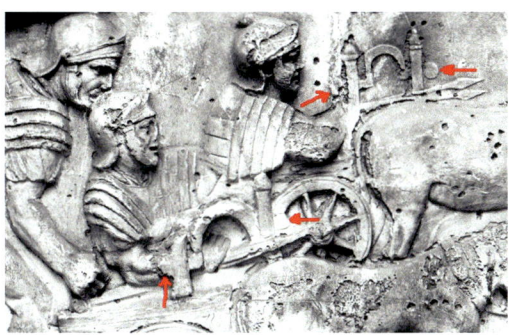 

Figure 96 - Trajan's Column: projections on the outside of the field-frame covers (photographs: the author)

The three arch strut catapults on the Column which have projections on the outside of the 'cans'. **(left)** Cast of the two *carroballistae* in Scene XL. **(right)** Arch strut catapult being transported on a cart in Scene LXVI Casts 163-4). The curved distortion is because I have combined the two separate sections of the E.U.R. casts.

On the *carroballistae* scene in Figure 96 the far machine is obscured by the back of a mule, but it has just released two bolts, which may be factually accurate, Zopyrus' *gastraphetes* (Chapter 2) and the Hatra catapults (Chapter 14) could shoot two missiles at once, or a mistake by the sculptor, or his way of suggesting a storm of missiles. The nearer machine is also in action, with the rear artilleryman's hand on the windlass handle at the end of the partly visible case, the drum being visible under his hand (there is a close-up in Figure 11). His colleague has his right hand touching what appears to be the slider, which is on a higher level than the case and projects beyond the arch strut frame. He may be loading a bolt onto the slider. In the scene of two *carroballistae* the rear legionary on the nearer machine is operating a winch handle. The size of a Roman cart, as estimated from the standard wheel spacing of five Roman feet, suggests that there was room in it for a larger version of the arch strut catapult than that in the *Cheiroballistra* text, perhaps with a rope spring diameter of 74 mm (3 Roman inches), as on the Lyon catapult (Figure 72).

The catapult manned by the two Dacians in Figure 97 must have been captured either during Trajan's campaign of AD 101-2, or some 15 years earlier when the Dacians had killed general Cornelius Fuscus and seized the catapults and engineers of Legio V Alaudae; Trajan recovered them during his campaign (Wilkins 1995, 8). If these two Dacians are in possession of one of Fuscus' machines, then the introduction of these arch strut catapults can be brought forward to *c.* AD 86-7. Note the object next to the bottom of the left-hand field-frame. I agree with Miks (Miks 2001, 210 Abb.32) that this looks like part of a winch with a central axle.

Figure 97 shows the four catapults with their forward projecting beams. Again, all four are in scenes of action. The sides of their projecting beams display the key detail of the male dovetails of the sliders. They are not cases, which have plain sides. It is significant that in the non-action scene of a catapult being transported on a cart (Figure 89) there is no projecting beam. Sliders only project when they are moved forward to slip the trigger over the bowstring at the start of the loading procedure.

Figure 97 - Trajan's Column: forward projecting beams – the sliders
Figure 85 **(top):** Scene LXVI Cast 164. Two catapults on a rampart next to a gate (see also Figure 35). **(below, left)** In the same Scene, Cast 166 depicts a catapult on a log emplacement. **(below, right)** Cast 169. Two Dacians with a catapult. All these catapults show the outline of the male sliding dovetail of the projecting sliders (photographs: the author).

Chapter 7

Deciphering the manuscripts: Vitruvius' *ballista*

The Vitruvian stone-thrower, the *ballista*, has been left to last because it is more complex than the bolt-shooter, and because there are no identified archaeological finds of parts. The catapult has to be reconstructed from the Latin text of Vitruvius, the description by Philon of an earlier version of the same machine, and some valuable diagrams and advice from Heron. Marsden has translated and discussed Philon and Heron's texts in Marsden 1971. Because Vitruvius was one of four officials appointed by Emperor Augustus to supervise manufacturing and repairing the Roman army's catapults, his description presumably represents the official version of the machine in use in the early Roman Empire.

As with all Greek and Roman texts, Vitruvius' text has suffered in transmission. Manuscripts, as the name implies, are hand copies. The earliest and best one of Vitruvius' *de architectura*, manuscript Harleian 2767 in the British Library, dates to *c.* AD 700, seven centuries after Vitruvius handed his master copy to his publisher in Rome. This is exceptionally early for a pagan classical ms, most of which usually date from the 10th to 11th century, or later. Harleian 2767 was written by monks in the same writing room as the Lindisfarne Gospels, possibly copied from one of the

Figure 98 - Len Morgan's two ballista *reconstructions*
(left) Len Morgan with his highly impressive Mk 1 two *librae* size Vitruvian *ballista*, based on the author's interpretation and model (Figure 107) (photograph: the author). **(right)** His Mk 2 version incorporates further adjustments, including a revised design of bowstring as a bowband (Figure 109) following Heron's description in *Bel*. 111, 3 – 112. (photograph: Margery Wilkins).

mss brought over from the continent by Benedict Biscop or St Wilfrid. Hand-copied mss can be traced back like human family trees; and just as a faulty gene may be inherited, so when a copy of a book is flawed, either because of a copyist's mistake or by physical damage, all copies descending from that one will be liable to repeat the flaw. There are clear signs that Vitruvius' text has garnered errors during the centuries of transmission. No diagram has survived; Harleian 2767 has blank pages for lost diagrams. Confusions have arisen because Vitruvius wrote both words and cardinal numerals in what we call Roman capital letters.

It is easy to forgive the mistakes made by copyists: a typical passage giving the length and width of a part would have been written: LONGITUDOFORAMINISSLATITUDOFORAMINISIS '*a length of half a hole, a width of one and a half*'. Both letter-numerals – S for a half, IS for one and a half – look the same as the end letter(s) of FORAMINIS '*of a hole*'. Most Latin manuscripts leave no or very small gaps between words.

Numerals like VIII or XIIII can easily lose digits. For fractions, engineers like Vitruvius used the first seven letters of the Greek alphabet. Even if the copyist did know the Greek alphabet, he would probably be ignorant of the obsolete digamma *F*, which can be mistaken for a Latin F or E, or a Greek gamma Γ. This letter continued to be used by engineers for the numeral 6 and the fraction 6/16 (see Figure 120 in Chapter 10 for its use on a *ballista* ball to mark six *librae*).

No classical pagan manuscript has been transmitted without errors, and the above explanation as to why and how a Latin text can be prone to error should not be used as an excuse to claim that Vitruvius' text is in such a state that it is useless as a description of the *ballista* and can be largely discarded. This is not so. One scholar (Campbell 2011, 687) talks about Vitruvius' '*badly mutilated Latin text*'. Aitor Iriarte too (2003, 127) is unhappy about the condition and therefore the reliability of the text. Both these views are justified by the treatment of the text in the most famous edition, published in 1917 by Schramm in collaboration with Hermann Diels; it contains his version of the Latin plus a commentary and interpretive plans. There is indeed an element of mutilation in this edition: Eric Marsden spotted that Schramm has included '*a number of items that would seem never to have been in Vitruvius' original*' (Marsden 1971, 198 and further comment on 200). My fresh study of the text (Wilkins and Morgan 2013) reveals that in dealing with the problems of the 49 numerals giving the sizes of *ballista* parts, Schramm has <u>without explanation</u> changed 20 to a reading quite unlike that written in the surviving mss. In almost all the 20 cases the existing mss readings make sense without requiring amendment. There are uncertainties in some passages, but Vitruvius' text is certainly not 'badly mutilated'.

The first numeral in the description of the *ballista* has been corrupted to a word, VEL meaning 'or'. The two strokes of the V may have been miscopied from the numeral II, and the E could originally have been F, digamma. The L can be explained as accidental repetition of the first letter of the following word, a common mistake because of the lack of gaps between words in most mss. This is the type of textual approach required to produce possible or, hopefully, probable answers to what Vitruvius originally wrote. The resulting measurement, two and six sixteenths, makes good sense in the context and unravels the rhombus figure described in the turgid Latin of Vitruvius' opening paragraph; encouragingly, the amended figure produces the same rhombus diagram as Heron's (Figure 100). From this rhombus the outline of the top and bottom hole-carriers can be drawn (Figure 99), an elegant geometric solution to the need to increase their strength where they have been weakened by drilling out the spring-holes.

The sizes of the components of the catapult were once again calculated in diameters of the rope spring /spring-hole. Following years of practical experiments the Greek engineers had devised the following formula for calculating this diameter, based on the weight of shot that the catapult was intended to throw: $D = 1.1 \sqrt[3]{(100 M)}$ where D = the diameter of the spring-hole/rope spring in dactyls and M = the weight of the proposed stone shot in Attic *minai*. The precision of the *ballista* design, worked out to cope with the variety of enormous stresses involved, is amply proved by this use of a decimal point and a cube root, reportedly the first known appearance of a third degree equation in the history of western mathematics.

The two half-spring frames

To project the weights of stone shot a *ballista* had to be capable of withstanding great stresses. Instead of the bolt-shooter's all-in-one spring-frame, each rope spring was given a separate, hefty spring-frame of strong plated timber (Figures 13 and 99). The two half-spring frames were clamped in place by a framework of top and bottom crossbeams and crossbars (Figures 102 and 105). Heron stresses the need to strengthen all critical points with iron plates applied to the sides of the stanchions of the half-spring frames, the outside of the top and bottom spring-frame components, and even the tenons of the stanchions inserted into the mortises in the hole-carriers.

The shape of the top and bottom hole-carriers was designed to place the stanchions out of line in *palintone* formation to allow an increased arc for the arms, and so increased spring power. Vitruvius and Heron describe the rhombus diagram which created the bowed outline of the hole-carriers, required in order to add extra wood to compensate for the weakness caused by drilling the spring-hole.

The loss of Vitruvius' diagrams makes the survival of Heron's doubly valuable. In Figure 100 copyists have stretched the figure horizontally: Heron states that the long sides, AB and ΓΔ, of his rectangle are twice the length of the short sides AΔ and BΓ. Heron's method of drawing the rhombus, based on this simple rectangle, is much clearer than Vitruvius', but if I have restored his measurements correctly they end up with the same figure. The main details of Heron's diagram can be redrawn to correct the horizontal distortion. The mss diagram gives the side-stanchions improbable and impractical slanting ends, presumably a copyists' error. I have squared off the ends and adjusted the mss lines HΘ and KΛ.

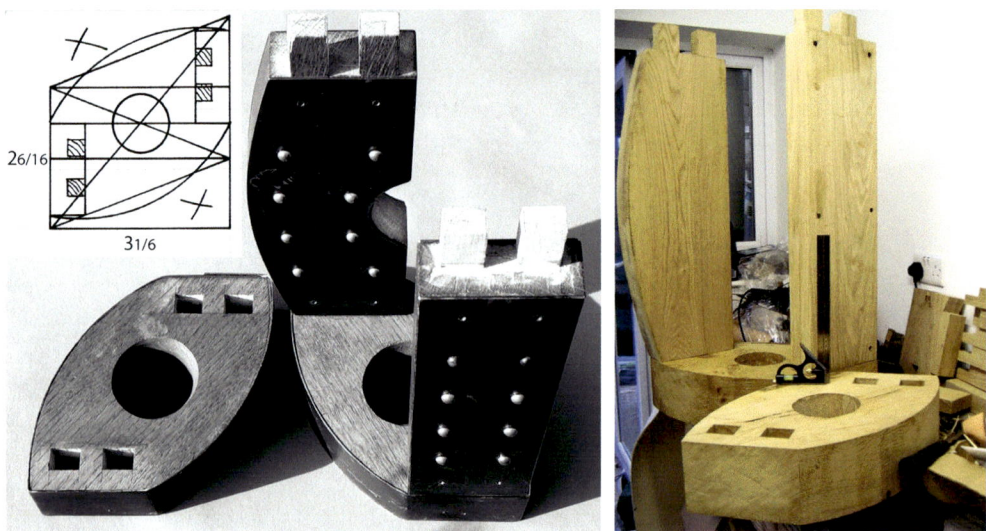

Figure 99 - Ballista half spring-frame
(left) The author's scale model of a half spring-frame, based on the interpretation of Vitruvius' rhombus figure at top left. The plating described by Heron has been added except on the tenons, which are not plated (photograph: the author). **(right)** Len Morgan's oak frame under construction to Vitruvius' two *librae* size. The semi-circular cut-out for the arm in the Side-stanchion has not yet been made (photograph: Len Morgan).

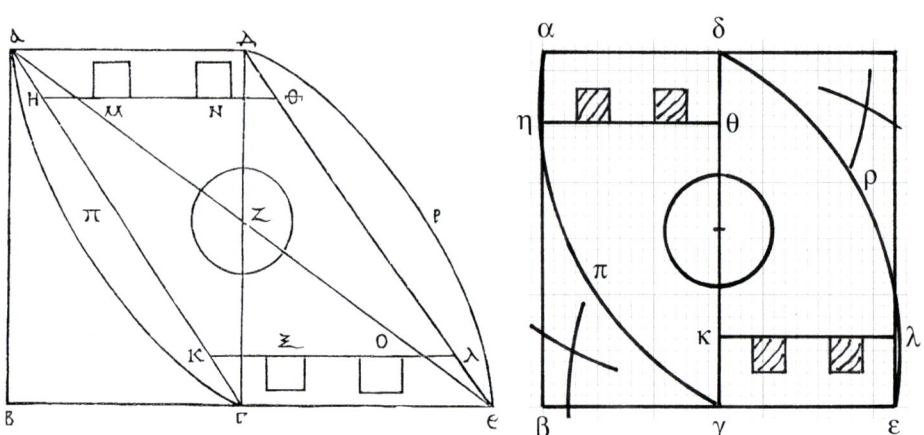

Figure 100 - Heron's rhombus diagram
(left) ms version of Heron's rhombus diagram (after Wescher). **(right)** The main details of Heron's diagram redrawn to correct the horizontal distortion.

The washers

The oval profile of the Vitruvius' washer's spring-hole (Figure 101 below) compensates for the space taken up by the washer bar and so allows vital extra spring-cord to be inserted. Catapult washers are usually circular and cast in bronze.

The sheer bulk of these Vitruvian washers on a large stone-thrower requires them to be made of plated wood, because huge bronze castings would be impractical. Len has used standard bronze washers for this, the smallest *ballista* on Vitruvius' list.

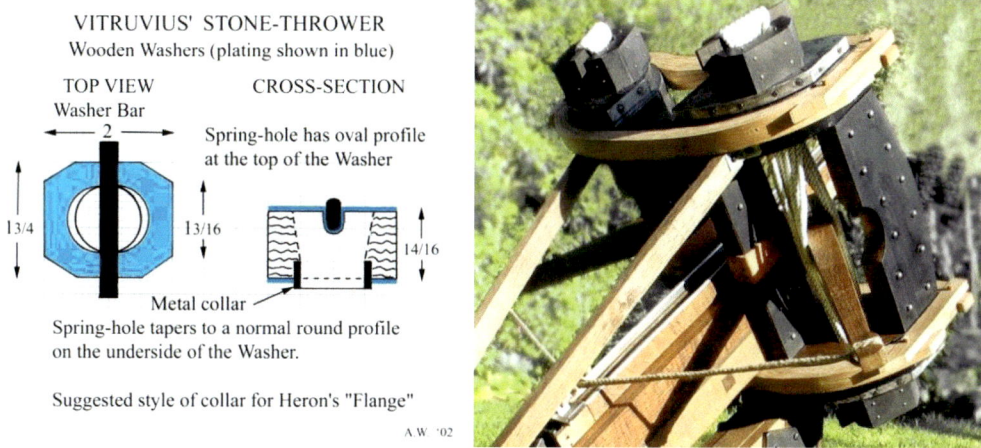

Figure 101 - Vitruvius' plated washers
(left) Vitruvius' plated wooden washer. Top and bottom plating shown in blue. The sides should also be plated. The wood grain should run vertically as on a butcher's block, and not as drawn. **(right)** Reconstruction of Vitruvius' washers on the author's model (photograph: the author)

The framework holding the half-springs and the 'inward-swinging arms' theory

The pair of half-springs is locked in position by an outer framework composed of the crossbeams and crossbars, identifiable in Heron's main diagram (Figure 102), although there the crossbeams are drawn as straight. Vitruvius clearly describes both as curved, as Marsden emphasised and used on the model made for him by Norman and Raymond Cooper. In spite of the ancient limitations in incorporating several planes in one drawing, and distortions from years of copying, this is a very helpful diagram. It contains two features not in Heron's text: a large horizontal bolt connecting the inner stanchions of the field-frames to the ladder, with reinforcing plates at the bolt's ends, and the four outer crossbars of the framework. Note the straight capstan bars and double winch ropes.

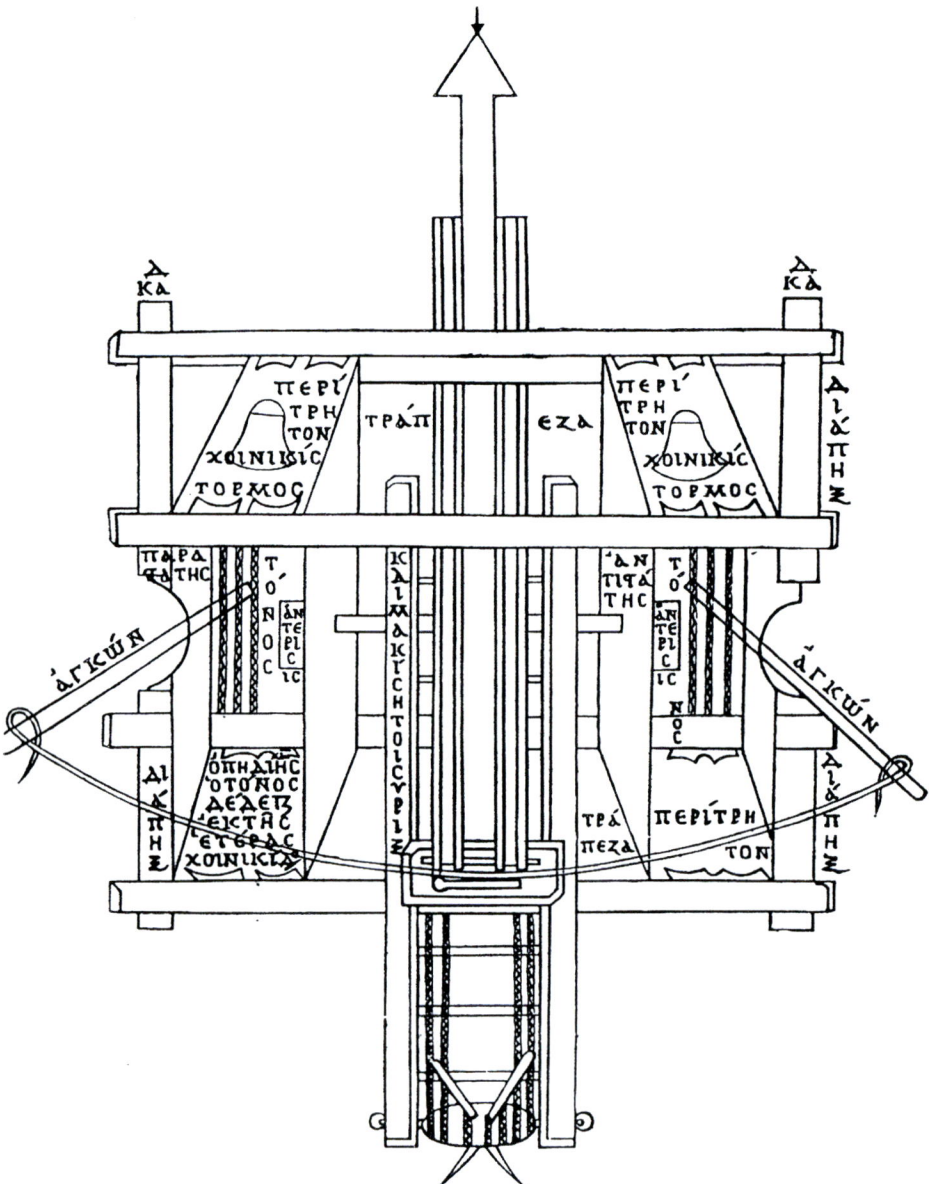

Figure 102 - Heron's ballista diagram
Heron's diagram of the *ballista* redrawn by Wescher, combining the figures and labels in the Paris and Vatican Libraries' copies.

On Marsden's model *ballista* (Figure 103), made by Norman and Raymond Cooper, halving joints are used to link the ends of the crossbeams and crossbars; this may have been Eric's final verdict on these joints. Note Eric's addition of extra crossbars inserted along the inside edges of the hole-carriers: such joints would seriously weaken the crossbeams at a critical point. In his diagram (1971 55, Figure 18) these

Figure 103 - Ballista model made for Eric Marsden (photograph: the author)

extra beams appear as ΧΨΦΩ; but these are the very letters which Heron's text (*Bel.* 99) gives for the rungs or cross-pieces of the table. Of course, Marsden was right to be worried about the centre sections of the crossbeams being under stress from the torsional forces applied to the half-springs when the arms are wound back and recoil. This model has now been returned to Eric's son Dr John Marsden and his family.

Figure 104 - Crossbeams
Len's crossbeams with top plating. The straightened central part of the front crossbeam, as on Marsden's model, helps to maintain the spacing between the two hole-carriers.
(photograph: Len Morgan)

Vitruvius begins by describing the rear crossbeam: *'The length of the beam which is on the Table is 8 holes, the thickness and width is half a hole. Tenons 5/8 h.[long], ¼ h. thick. Curvature of beam 19/16 h.'*. Philon gives the crossbeams' width as 5/9 h., their thickness 4/9 h..

Vitruvius describes the front crossbeam: *'its width and thickness are the same* [as the rear crossbeam]; *its length is what the actual curve of the design* [of the Hole-carrier] *and the width of the side-stanchion add to its curvature'*. This appears to mean that the beam is shaped to follow the contours of the front edges of the Hole-carriers (Figures 104 and 105).

The ends of the crossbeams are linked by crossbars which the three engineers do not mention in their texts, but all four bars are shown clearly in Heron's diagram, where they are labelled *diapex*. They sit flush with the outside edges of the Hole-carriers and link the ends of the crossbeams.

Whereas the bolt-shooters had *euthytone* frames where the stanchions are in line, as on the Xanten-Wardt *scorpio* (Figures 36 and 37), the Figure 105 diagram shows how the *ballista*'s staggered *palintone* position of stanchions allowed a greater arc of arm travel. The bowstring, says Heron (*Bel.* 110), *'takes the shock of discharge'* and (*Bel.*

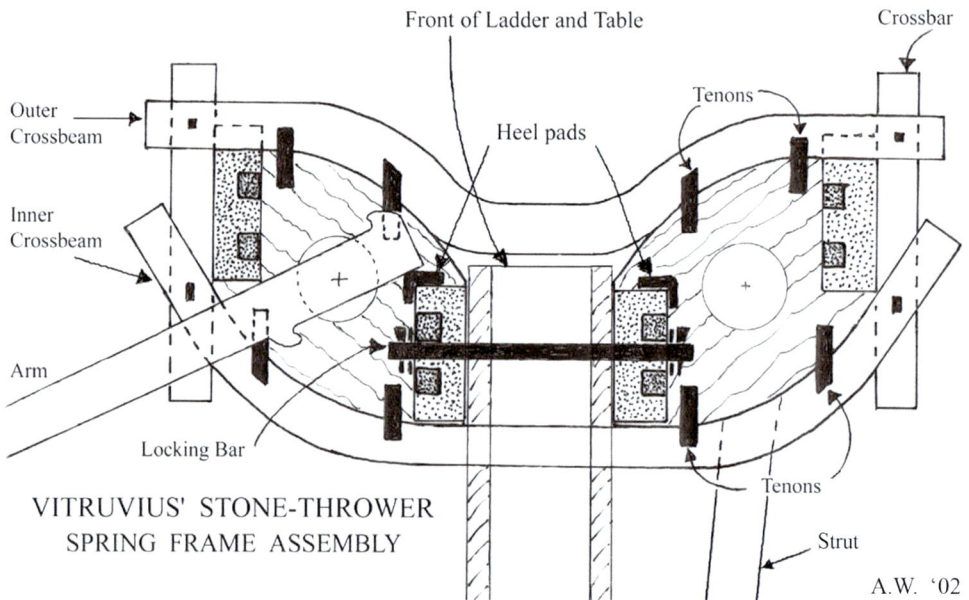

Figure 105 - *Diagram of the* ballista *framework*
My original drawing of the framework and locking bar from above, along with the positions of the ladder, the right-hand stay/strut and left-hand arm. I have extended the crossbars and crossbeams far too much beyond their halving joints.

102) *'it is necessary to tighten the bowstring sufficiently to hold the arms slightly clear of the side-stanchions to prevent them crashing into them and inflicting and receiving damage.'* The heel pads on the front of the centre-stanchions are part of this braking system. The damage to the side-stanchions of the BBC's one talent *ballista* by the arms (page 126) happened in this way, because the bowstring stretched when the first shot was launched, the correct prestretched rope having failed to arrive in time.

Some writers (e.g. Iriarte 2003, Campbell 2011) have discussed the theory that *palintone* catapults such as the arch strut bolt-shooters, may have had inward swinging arms which move in an arc from behind to the front of the spring framework. They have looked for clues that the *palintone* stone-throwing *ballista* might also have had its half-spring frames spaced sufficiently wide apart to allow inward swinging arms. First, Heron's diagram (Figure 102) indisputably shows outward swinging arms, and he states (*Bel.* 100.2) that the arms recoil *outwards* (*eis to ektos*); the Greek for *forwards* would have been *eis to prosthen*. Second, these writers have looked at Vitruvius' description of the *regulae* (crossbeams) in the hope that they could be lengthened far beyond his figure of 8 holes. Campbell (2011, 687) claims that 'the crucial passage where Vitruvius describes the *regulae* (rods) connecting the spring-frames is hopelessly garbled'.

That is not so: there is no problem whatsoever with the Latin text at this point: *regulae quae est in mensa longitudo foraminum **VIII**, latitudo et crassitudo dimidium foraminis*. 'The length of the beam which is on the Table is **8 holes**, the thickness and width is half a hole'. Looking at the problem from an engineering point of view, if the *regulae* were to be extended at all to, say, **X**VIII, 18 f. to allow the Half-springs to be spread out to create inward swinging, they would, with Vitruvius' '*dimidium foraminis*' (half a hole) cross-section, be liable to break. Note that their cross-section in Philon's earlier version of the machine is only 5/9 h. wide by 4/9 h. thick. Interestingly, the Morgan-Wilkins reconstruction of the Hatra catapult (Chapter 14 and Figure 153), does interpret it as a spread-out version of the Vitruvian *ballista*.

As it stands, Vitruvius' 8 h. length is just right to create a framework that clamps the two half-springs to the sides of the ladder. The framework is finally locked there by the bar on Heron's diagram, which I have labelled the 'locking bar': it passes through the ladder and the two inner stanchions of the half-springs, with the two reinforcing plates/blocks labelled *anterisis* at its ends (Figure 102). This is exactly the same system as that on the Xanten-Wardt *scorpio*, where two long rivets pass through the case and through the inner stanchions; the rivet ends are peened over two rectangular plates on the outside of the centre-stanchion (Figure 37). On our reconstruction Len has put a second bar through the table (Figure 106).

Figure 106 - The crossbars and the table
(left) Details on Len's Mk I machine of one of the plated crossbars joined to the ends of the crossbeams by halving joints and pins (photograph: the author). **(right)** Len Morgan's Table, with the plank ready for fastening on top of the sides and rungs (photograph: Len Morgan).

On Len's Mk 1 *ballista* (Figure 106) two small metal straps are additions by Len to secure the crossbeams to the hole-carriers; these modern additions were found to be unnecessary on Len's Mk II version shown in Figure 98. As mentioned in the text above, we have used two locking bars, the bolted ends of which can be seen behind the rope-spring, though we have not added the reinforcing plate, *anterisis*, included on Heron's diagram.

The Ladder and the Table

The case is not made out of a solid beam of timber, which would be impossibly large and heavy, but is built as a ladder, with broad side-poles and rungs. The ladder rests on top of the table (Figure 107) which in reality is a second ladder with a cover like a table top. The table is needed to lift the ladder so that the slider holds the stone missile in a centre alignment with the bowstring. This arrangement and the ends of the rungs are visible on Len's full-size *ballista* and on my model in Figure 107.

The trigger

The trigger for the *ballista* only has one 'finger' (Figure 108), which Heron says (*Bel.* 111, 5-6) fits into the ring on the back of the bowstring (Figure 109).

Bowstring problems

Heron (*Bel.* 111, 3 – 112, 2) says that the stone-thrower's bowstring is not round like the bolt-shooters, but flat like a belt with a sort of ring in the middle woven from the sinews of which it is constructed (Figure 109). Into this ring fits the single finger of the

Figure 107 - The author's model ballista
The author's large-scale model, with the curved *regulae* (crossbeams) of the framework and the special plated wooden washers described in Vitruvius' text. I have added Philon's counterplates (Philon 53.16) under the washers; they are not in Vitruvius' text and are omitted on our full-size and quarter-scale reconstructions. The bowstring should be much wider (photograph: the author).

trigger. The bowstring is positioned so that it catches the middle of the stone missile. '*If it is positioned slightly too high or low it will either travel under the stone or slip over it.*' The Kaiser was almost a victim of the *ballista* when Schramm's bowstring slipped under the missile, propelling it upwards instead of forwards (Figure 30).

Figure 108 - Trigger and mounting plate
(**left**) Looking down between the two stays onto the trigger, mounted on the rear of the slider. Its release lever is in the released position, having been operated by a long rope. (**right**) Side view of Len Morgan's trigger and mounting plate. (photographs: Len Morgan)

Initial attempts to make the bowstring out of two ropes, only filled in like a band in the centre, as we did with the large BBC *ballista*, allowed the bowstring to wobble and miss the centre of the missile. Len's impressive final version is fully belt-like and functions very reliably. Philon (*Bel.* 54.1) gives the length of the bowstring as 2.1 times the length of one arm, a measurement which on its own rules out the possibility that this is a wide-frame catapult. This is exactly the measurement of Len's bowstring: 73½ inches long for an arm 35 inches long. I only asked Len to check this measurement several months after the catapult was built – we have not engineered this result, which in itself lends some support to the general accuracy of our reconstruction.

Figure 109 - The bowstring belt
Len's final version of the bowstring belt, following Heron's description. The single finger of the trigger (Figure 108) fits into the ring at the centre of the bowstring belt (photograph: Len Morgan).

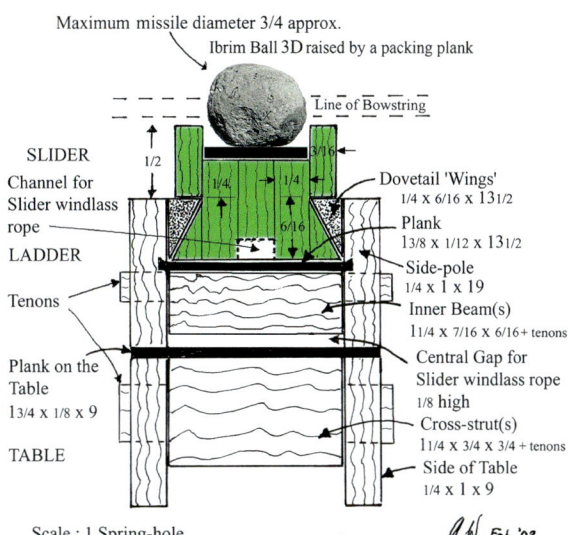

Figure 110 - Cross-section of table, ladder and slider
(left) Realisation by Len of the author's interpretation **(right)** of Vitruvius' description of the cross-section of table, ladder and slider (in green). (photograph: the author)

In Figure 110 note the bracket and spindle holding the pulley wheel feeding the winch rope from the central gap under the Ladder's inner beams to the channel in the base of the Slider. This allows the heavy Slider on large *ballistae* to be winched forward up the inclined Ladder to connect the trigger to the bowstring. For the Qasr Ibrim missiles see Chapter 10.

'Building the Impossible': the BBC *ballista*

The BBC approached the author for help in constructing a very large version of the *ballista*. Throughout Greek and Roman history huge stone-throwers projecting three talent (78.6kg) stones seem to have been used only on rare occasions, and usually by a besieging army; few city walls could have accommodated even a half-talent (13kg) stone-thrower, and very few the one talent (26.2kg) machine which we reconstructed. We chose the one talent size because it appears to have been the largest in regular use by the Roman legions, and because the catapult engineer Philon of Byzantium describes it as '*the most violent*' stone-thrower, and gives detailed instructions for constructing a triple ditch system round a city to keep this deadly catapult sufficiently far away to reduce its impact on the city walls (Figure 114). The siege of Jerusalem was chosen because a graphic eyewitness description of the sight, sound and impact of the one talent shot is given by the captured Jewish general Josephus in his *History of the Jewish Wars*. Arbiv 2023 reports the discovery of a 24kg *ballista* stone from this siege. See Chapter 8.

Constraints of time meant that the BBC only had three hours to shoot the reconstruction, enough for only three shots; and that there was no time to cope with a bowstring problem: it stretched so much after the first two shots that the arms smashed into the spring-frames' unplated side-stanchions, causing fatal cracks. To be fair to the BBC, the correct type of <u>prestretched</u> rope for the bowstring failed to arrive. The bowstring, says Heron (*Bel.* 110), '*takes the shock of discharge*' and (*Bel.* 102) '*it is necessary to tighten the bowstring sufficiently to hold the arms slightly clear of the side-stanchions to prevent them crashing into them and inflicting and receiving damage.*'

See also the photo of the BBC *ballista* launching the final stone in Figure 142 in Chapter 12.

The eight-man crew of the BBC *ballista* (Figure 111), perhaps numerically the same as their Roman counterparts, is winching the slider forward so that the trigger can be locked onto the bowstring. Constraints of time and money, and changes imposed by the BBC's consultant engineer, meant that there are several unauthentic details:

Figure 111 - The BBC one talent Vitruvian ballista *(photograph: the author)*

the stand is based on the two-column version invented by Schramm and Diels. The 45° angle of elevation is too high for shooting to destroy city walls, and would only be required for reaching the extreme height of the fortress at Masada (Figure 116). Like our own first attempt with the Morgan-Wilkins *ballista*, the bowstring was made out of two ropes, only filled in like a band in the centre; it allowed the bowstring to wobble and miss the centre of the missile. Our attempt at building this one talent stone-thrower left us in awe of the massive and complex engineering which coped with compressional forces measured, according to the BBC engineer, in hundreds of tons, and it confirmed our respect for the Roman legions' mastery of the supply and construction of huge volumes of timber, metal and sinew-cord. *Ballistarii* ('artillerymen') were among the specialist soldiers in a legion who were excused fatigues such as guard duty. The erection of such a monster machine must have spread terror amongst the defenders of a city, as did Edward the First's 'War Wolf' trebuchet, which so terrified the defenders of Stirling that they offered to surrender without a fight.

The energy remaining after the missile leaves the bowstring is absorbed partly by the bowstring and partly by the arms and the spring-cord, as the inner ends of the arms strike the heel-pads attached to the inner-stanchions (Figure 105).

The arms

Tests on the BBC *ballista*, on my large model, and recently on Len's two *librae* machines, found that an arm six spring-holes long was too short; fitting an eight holes long arm with its longer bowstring would place the trigger position so far back that it would leave inadequate room for the winch and pulley system. An arm seven holes long, as on the scorpion bolt-shooters, works well. The bolt-shooter's spring-frame is *euthytone* not *palintone*, i.e. the stanchions are not offset, so that curving its arms is the only way to increase arm travel. The *ballista* arms are straight, because the increased arm travel is achieved by the staggered *palintone* position of the stanchions of the spring-frames. Philon's arms had a square cross-section, so Vitruvius' presumably did.

Confirmation that we have the correct arm length comes from Philon's statement (*Bel.* 54.1) that the length of the bowstring was 2.1 times the length of one arm. As already mentioned above, this is exactly the measurement of Len's bowstring: 73 ½ inches long for an arm 35 inches long.

The stand and the angle of elevation

The design of the stand of the BBC catapult was based on the unauthentic twin-column version invented by Schramm and Diels, who altered the manuscript reading. In fact, Vitruvius lists a single support column for the stand with a width and thickness of 1½ spring-holes, as against the bolt-shooter's 3/4h. There is a good case for believing that Vitruvius' stand was similar to that shown on the Cupid Gem (Figure 85) and

Figure 112 - The Morgan-Wilkins two librae *catapult in action*

At the 2019 Durham Roman Army Conference the Roman Military Research Society assembled and shot the *ballista* on the site of *Viminacium* (Binchester). For safety reasons we always avoid stone missiles and hold the world record for launching the fastest melons. (photograph: Mike Haxell)

Trajan's Column (Figure 57), i.e. a single hefty column braced by two diagonal struts.

Vitruvius says that the height of the stand should be '*as practicality dictates*'. Although an elevation of 45° is a suggested textbook angle for achieving maximum range, field tests by the Roman Military Research Society with our three-span *scorpio* found that the bolts shot at 35° travelled 6-7% further than those from an elevation of 45°. The BBC ballista should have been angled at 35° or lower for attacking a city wall. The 25° elevation of the Cupid Gem case may be close to the authentic angle of elevation for shooting against advancing enemy troops.

Chapter 8

The stone missiles: range and effects

For the effects of the heaviest stone shot there are the more credible parts of the eyewitness description by Josephus (*Jewish War* III, 243) who had been on the receiving end of these missiles at the siege of Jotapata in AD 69: '*...more terrifying than the machines themselves was the whizzing sound of the stones*' which '*tore off the battlements and broke off the corners of the towers*'. This is reminiscent of the First World War descriptions of the sounds of shells passing over. '*There is no body of men strong enough to avoid being flattened right through to the back rank by their force and size.*' This may be explained as the bouncing bomb effect of stones ricocheting on hard surfaces and causing casualties right to the end of their travel. The stones are multi-strike missiles; in contrast catapult bolts are only likely to be effective on first striking a target. In AD 70 Josephus watched the siege of Jerusalem from the Roman lines, praising the catapults built by the Tenth Legion as superior. '*Their scorpions were more powerful and their stone-throwers larger. The stones which they threw weighed one talent and travelled two stades or more.*' Two stades (368m) is difficult to believe, and is usually dismissed as an exaggeration typical of Josephus and/or Roman army propaganda. However, it might just be accurate for the range of the 26kg shot at the end of its travel after numerous bounces. The rest of Josephus' account sounds authentic: '*...warning of the approach of the stone missiles was given not only by their whizzing sound but also by their whiteness. So sentries were posted on the towers who gave warning whenever the machine was fired and the stone was on the way by shouting in their native tongue 'Sonny's coming'*. Those in the path of

Figure 113 - Cutting a one talent stone ball
Distinguished professional sculptress Carole Kirsopp cutting a one talent (26.2 kg) limestone ball for the BBC catapult (photographs: the author).

the missile then scattered and lay flat, the result of their precautions was that the stone passed over without causing casualties'. The Roman answer was to blacken the stones, almost certainly carved from the white limestone of the Mount of Olives; *'they could not be seen so easily and hit their target, killing many with a single shot.'*

The process of cutting a *ballista* stone (Figure 113) begins with mathematical calculations based on the weight of the stone per cubic metre. This gives the required size of cube onto which is scribed a cylinder (centre photograph). The cylinder is then cut into a sphere, one half at a time, a special box (right photograph) being used to support the half-trimmed sphere. The process took Carole two and a half days. The result was as spherically precise as the balls from Carthage in Figure 14. Arbiv 2023 reports the discovery of a 24kg ballista ball from this siege, only a fraction lighter than Vitruvius' weight for a one talent missile.

So, one talent catapults were lethal against personnel and wooden or lightly built stone battlements, but had no significant effect on the walls of Jerusalem, which Herod the Great had rebuilt to the highest specifications with huge blocks of stone.

A vital and unique piece of evidence on the one talent's effective range comes from Philon's work *Poliorketika* ('Siegecraft') dealing specifically with a triple ditch system to reduce the damage to a city wall by a one-talent machine *'which is the most violent'* (Figure 114 below). He is talking about a strong defensive wall built of large blocks of stone. The third of his deep, well-spaced ditches will keep the catapult about 163m from the wall when it *'will strike it with reduced/weakened force'*. The space between the third and second ditches is to be denied to the catapult by a 'minefield' of thorns, stakes and ditches, the implication being that if the enemy can station the catapult there at about 113m from the wall it will seriously damage it. From this it can be argued that at a range of 163m the stone was losing the power to do serious damage to a wall of large stones because it had travelled well beyond the point at which it first struck the ground. If the underlining assumption behind Philon's statement is that the wall will only be damaged by a direct hit, this would mean that the usual point of first ground strike for this size of stone-thrower was 113m or beyond. If this is so, then the evidence of Philon's distance measurements in the diagram below could mean that the first bounce of the one talent stones would be in Ditch 1; this might prevent the stone from bouncing at all and so going any further.

This is the performance of the Hellenistic version of the machine; but Vitruvius' washer design allows more spring-cord to be inserted, and the revised text reading (Chapter 7. 2) shows that the spring-frame was enlarged to allow room for this.

Preliminary tests with Len Morgan's Mk I two *librae* machine achieved first strike ranges of around 120 to 130m with slightly overweight stones, but this smallest size of *ballista* should probably be reaching further, and we have not yet extracted the maximum performance from it.

The stone missiles: range and effects

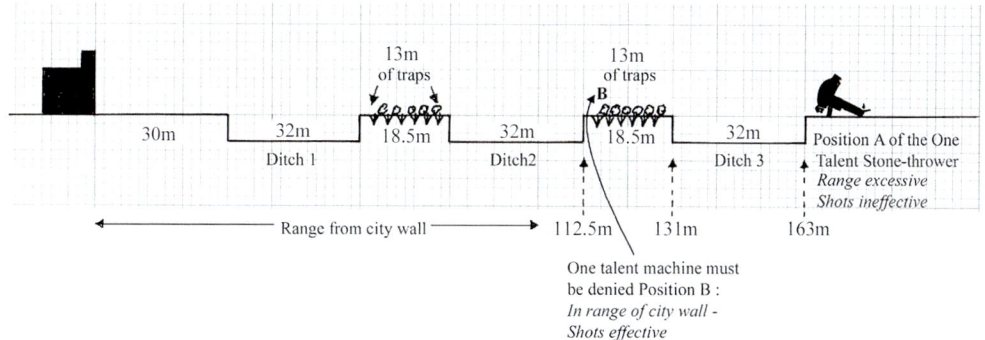

Figure 114 - Philon's defence against the one talent ballista
'When ditches of this size and design have been dug, it will be impossible to fill them in quickly, and the one talent stone-thrower, which is the most violent, will either fail to get close to the wall, or will strike it with spent force.' Figure 114 is drawn with acknowledgements to Eric Marsden (1969, 90-91 and Figure 2). Note that I differ from Eric in the translation of the last phrase.

A *ballista* and sling 'target' on the Rhine frontier

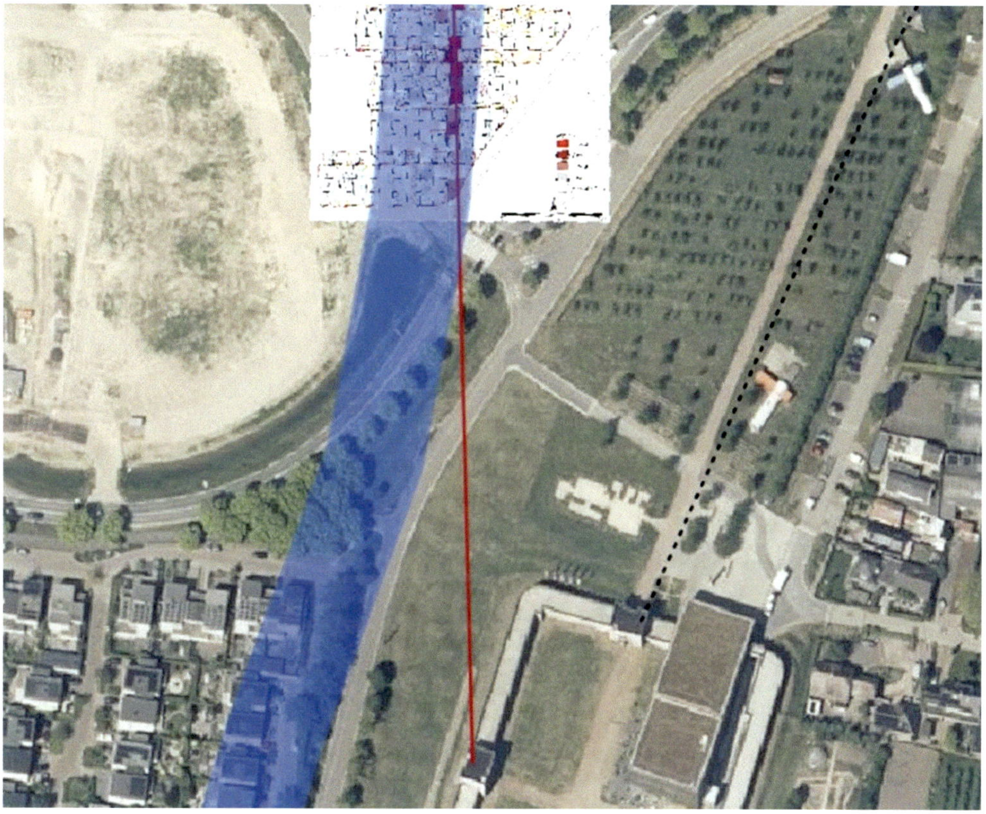

Figure 115 - Ballista *stones and sling shot in a long line on the former river bed close to the De Meern fort.*

From excavations on the Rhine frontier at De Meern near Utrecht in the Netherlands (see Aarts A. C. 2012), has come a very remarkable discovery of Roman artillery in action, in the form of a long line of catapult balls and slingstones in the former bed of the Rhine. In the top of Figure 115 the red squares mark the sections of the river bed where the missiles were excavated. See Figure 3 for an enlarged view of these squares. Erik Graafstal has traced a line through these squares back to the possible launch position at the *porta praetoria*, the west gate of the fort. The distances of the line of missiles from this gate range from 167m to 230m, and the weights of the stones range up to 8.21kg. It is improbable that hand-thrown stones could reach 167m, but this distance might be achievable with the help of a *fustibalus* throwing stick. The lighter stones in this discovery are too small for a *ballista*: Vitruvius' list begins with the two *librae* catapult's stone weighing 0.65kg. So, these were launched by slingers.

A catapult ball weighing 7.24kg has reached the archaeologists' square 683 at a range of about 225m. The first bounce range of a Vitruvian *ballista* seems likely to be up to 180m. It seems almost certain that the De Meern garrison was using the wide-frame Hatra type of stone-thrower (Chapter 14) which many believe was the Roman army's upgraded design as a parallel exercise to that of the wide-frame arch strut bolt-shooter. Tom and Len's superb scale models of the Vitruvian and Hatra catapults (Figure 153), made to an identical scale, achieved 35.7m (Vitruvian) and 52.53m (Hatra) with a 94g projectile. The Hatra shot half as far again as the Vitruvian. This De Meern evidence makes it likely that the Hatra design of stone-thrower was available to the legions in the west.

The extreme angle and long range of shooting from the fort strongly suggest that this is in response to a real threat rather than shooting practice. The directional accuracy of the shooting is very impressive. The target is parallel to and close to the south bank of the river. The varying range of shots could be evidence that the target was moving.

In summary, this incident can be claimed as the third known 'target' of Roman artillery, along with Sulla's damage to the walls of Pompeii (Chapter 3 and Figures 15 and 16) and the attack on the chieftain's hut at Hod Hill (Chapter 4 and Figure 22). The De Meern balls are dated from the mid second to the mid third century AD.

Erik Graafstal has added the following information: 'Perhaps an additional point: we have evidence of artillery (*ballista* balls or bolts) from all four excavated small installations (probably all of them watchtowers) west of De Meern, two of them dating *c*. 40-70, two between 130 and 230. In the former case, I would not rule out the possibility of mixed (i.e. part-legionary) occupation of the installations in the Rhine delta, but in the latter I would. The least it shows is the high resolution of deployment of artillery in Roman frontier security systems.'

Chapter 9

Masada AD 73-74

The Tenth Legion ended the war against the Jews by besieging the final resistance fighters in Herod's vertiginous rock fortress of Masada. The giant siege ramp built upwards along an existing spine of rock, and the hauling of the 27m high, iron plated siege tower up it, are amongst the great feats of engineering in the history of siege warfare.

The Figure 116 photograph of the 1966 Masada exhibition shows Eric Marsden's three-span *scorpio* (Figures 12 and 21). The group of stone shot on the left are of large orange or grapefruit size, probably about six *librae* 1.96kg. The left hand one of the three large stones is probably a 26kg one talent shot, the centre and right-hand ones *c.* 21kg. The enormous stone on the right is explained by the excavator Yigael Yadin as one of the 'rolling stones' prepared by the Jewish defenders for sending downhill onto the Roman siegeworks. Josephus (Jewish War IV, 75) records that the Jews

Figure 116 - Stone missiles at the Masada Exhibition (photograph: Eric Marsden)

defending Gamla in AD 69 had caused havoc 'in particular by rolling down stones' on Vespasian's advancing troops. The tiny modern figures in the wall-filling exhibition photograph on the right show the enormous scale of the Roman ramp up which the siege tower was hauled. The possible site for a one talent or larger ballista, suggested by the author (text below), is marked with an X.

Yadin's excavations of the rock top uncovered hundreds of stone missiles, but I can find no mention of catapult boltheads. I suggest that bolts would have been a less effective type of ammunition during the main months of the siege because of the unique problems posed by the great height of the fortress above the surrounding terrain and by other factors such as the obstacle of the casemate wall. The only way to land stone or bolt missiles onto the top was to angle the catapults at 45 degrees or steeper and launch the projectiles high in the air so that they dropped down onto the interior of the fortress. Bolts shot in this way would start their descent at a fairly slow velocity and would have done little or no damage to structures because of their light weight (85gm for a two span scorpion missile, up to180gm for a three-span scorpion, about 300gm for a two cubit catapult). Also, because they were single-impact missiles with a very small cross-section and by the end of their travel were not accurately aimed, their chances of hitting the small target presented by a person would be very slim. If the Romans did try dropping bolts onto Masada the Jews would have welcomed the iron boltheads and turned them into arrow-heads. This may have happened.

In contrast, Roman stone missiles were vastly heavier, up to 26kg or more, had a wide profile and different impact characteristics. The description by Josephus of the effects of stone-throwers at the siege of Jerusalem, particularly those of the Tenth Legion, (Chapter 8, opening paragraph) records the fearsome reputation that these machines had already earned. The Legion's Masada stone missiles appear to belong to varying sizes of stone-thrower from two to 80 *librae* calibre (Holley 1994; see Vitruvius' list in *Weights and Measures*). One of the 26kg (80 *librae*, one talent) stones was exhibited in the Masada Exhibition of 1966 (Figure 116). The gravitational acceleration of the heavier stones will certainly puncture roofs, and if they hit solid objects like walls or pavements they will bounce and/or ricochet and may split to become multi-impact missiles lethal to personnel. It is of course possible that to clear the height of the rock and the casemate wall, 4 *librae* machines were using two *librae* ammunition, 16 *librae* catapults shot 8 *librae*, and so on. The fact that 26kg stones have been found on the top (Figure 113 for their manufacture) demands an explanation. In view of the difficulties posed by the 1 in 5 slope of the siege ramp it seems less likely that the Romans would have followed Demetrius Poliorcetes' idea and mounted the extra weight of a one-talent catapult in their 27m high iron plated siege-tower. The BBC one talent *ballista* weighed ten tonnes. However, if a one talent machine was positioned at the front end of the White Rock, at point X in Figure 116, and elevated at about 50+°, it could, in the author's opinion, have sent stones soaring high over the casemate wall and dropping down on Herod's buildings. If it failed to do so, the answer would have been to use a one-and-a-half talent or larger machine to fire one talent stones (using a spacer in the slider to cope with the undersize shot (Figure 110). The arc and impact of the missiles could have been observed and corrected from the key position of the Roman Camp H, which looked down onto the fortress top.

According to Josephus (*Jewish War* VII, 309) bolt-shooters were used against the fortress, mounted in the siege tower for the final assault. He says that volleys of missiles from many scorpions and *ballistae* drove the Jews off the ramparts. Again, no boltheads are reported from the excavations. Josephus was not there, so he may have simply assumed that this standard use of bolt-shooters did take place. It would be remarkable if the Romans had succeeded in finding and collecting every single one of the boltheads after the end of the siege.

I offer another possible explanation: that the Tenth Legion's supply of bolt-shooters was used up in deployment on the ramparts of the eight encircling forts and also on the 15 towers on the wall of circumvallation, ready to destroy any Jewish attempt to break out, and with fresh memories of the Jews' dangerous attacking sallies against the walls of Jerusalem. It could be that only *ballistae* were used to bombard Masada, and that archers and *ballistae* were used for the final assault from the siege tower. This tallies with the archaeological evidence of missiles on the top of Masada.

In contrast many iron catapult boltheads have been found in excavations on the spectacular site of the Roman siege of Gamla in AD 67; some 2000 *ballista* balls have been found there, many apparently *in situ*. (Holley 2014; Campbell 2006, 167 and photograph on 170). The siege is described by Josephus IV, 1-83. For archaeological results from the siege of Jerusalem see Arbiv 2023, reporting *ballista* stones of 0.33, 2, 5, 8, 11 and 24kg. 24kg is fractionally under Vitruvius' weight for a one talent *ballista*.

Chapter 10

Qasr Ibrim: artillery in defence. Inscribed stone shot

Contemporary with the publication of Vitruvius' treatise *c.* 25 BC, a large group of *ballista* balls from Egypt, 38 with carbon ink writing, has produced a vivid picture of a small Roman legionary garrison, isolated in enemy territory, using stone-throwers in a defensive role. The unique ink inscriptions tell a story of rivalry between the five centurions, and the dominance of one. As noted above (Figure 16 bottom right), the bulky and heavy stone ammunition for *ballistae* had to be cut on the spot by the legionaries. However, because of the friable nature of the local sandstone, half the Qasr Ibrim balls split during manufacture. The finds of large numbers of wine *amphorae* may be interpreted as a means of compensating for the great frustration and anger of the stone cutting crews.

Before the creation of Lake Nasser, the spectacular fortified site of Qasr Ibrim, south of Abu Simbel, towered above the Nile (Figure 117).

Figure 117 - Qasr Ibrim before the Aswan Dam
A last view of Qasr Ibrim, taken in the 1960s before the completion of the Aswan Dam and the rise of Lake Nasser (photograph: McDonald Institute, Cambridge).

Qasr Ibrim: artillery in defence. Inscribed stone shot

Strabo (*Geography*, 17.1.53) records the events of 25/4 BC when the local Meroitic Ethiopians, furious at their treatment by the Roman tax collectors, attacked the three cohort Roman garrison at nearby Syene (Aswan) (Figure 118). In retaliation the Roman governor Petronius routed their army, capturing the Meroitic generals. He lay siege to Premnis (Qasr Ibrim) and left a garrison of 400 legionaries there with bolt-shooting and stone-throwing artillery and food for two years. The 1978 and later excavations at the site for the Egypt Exploration Society (London) found over 780 *ballista* balls and the tanged boltheads and foreshaft (Figure 50).

Among the Hellenistic and Roman caches of stone ammunition discovered to date, many stones are inscribed or painted with their weight. The ink writing on the 38 Ibrim balls is unique in that it contains much more information than just the weight. Several stones are ascribed to particular centurions, whose striking names, Pompeius, Iulius, Octavius and Antonius, raise the suspicion that they have appropriated these famous names rather than received them by formal grants of citizenship. One ball, with the initials M.V, probably the senior centurion, perhaps Marcus Vipsanius (i.e. Agrippa) is even labelled as a missile intended for the enemy queen. The fact that the weights are recorded in both Greek and Latin script makes it likely that that the troops involved were from one of the legions originating from the Greek-speaking eastern Mediterranean and stationed in Egypt in 23 BC. The stones are the earliest to be discovered with their weights recorded in Roman *librae*, rather than the Greek *minai* marked on earlier finds of stone shot. This is evidence of the imposition of the Roman army system in Egypt, which was now under the strict personal control of Augustus.

In Figure 118 right the areas where clusters of ballista balls were found are shaded in grey: see Figures 119 and 122 below. WS = water stairs down to the Nile. NWB = north west bastion. PG = postern gate. EG = eastern gateway.

Figure 119 shows a cluster of some 34 *ballista* balls in South Rampart Street (see the position on the plan in Figure 118). If this is not a random group, but the ammunition for one catapult, then the variation in the size of the missiles backs up the evidence from the inscribed balls that a *ballista* was used to hurl balls smaller than the maximum diameter for which it was designed. The scale ruler divisions are 10 cm. See Figure 110 for the design of the launching channel.

Figure 120 shows **(top left)** Ball 2H. Diameter ±12cm. The vertical arrow appears to be the Greek sign for weight, perhaps standing for the point of balance of weighing scales. ⌐ is a *digamma*, the sixth letter of the very early Greek alphabet. It only survived as the numeral letter for 6 or 6/16ths. The second line repeats the weight in Latin script: P[onderis] VI 'of six [*librae*] weight'. That the weight is in Roman *librae*, not Greek *minai*, is proved by the two balls in the British Museum, 3E and 3F. In line three ⌐ is the sign for centuria, commonly cut as a reversed capital C on building

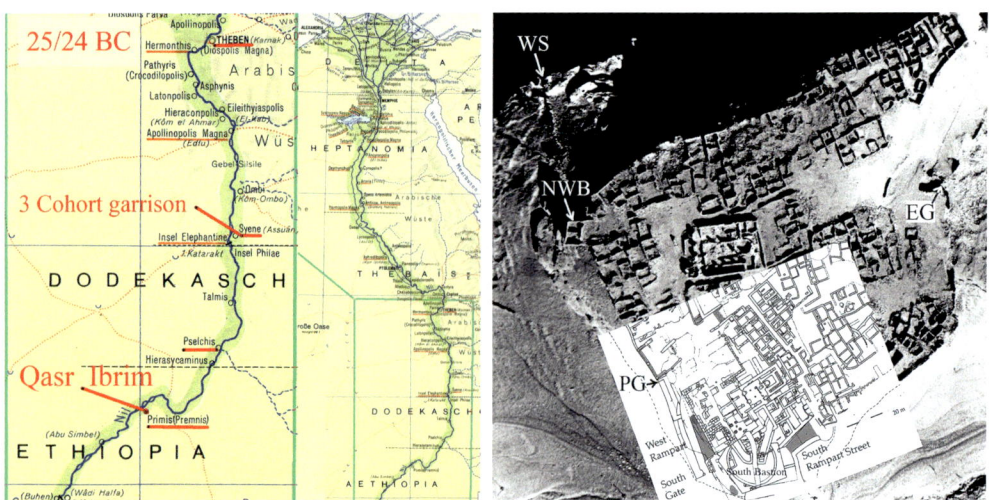

Figure 118 - Map of the Nile. Airphoto of the site
(left) The Nile area from Thebes to Abu Simbel **(right)** Aerial view of Qasr Ibrim with the Nile showing black, and the excavated area superimposed.

Figure 119 - Ballista balls in South Rampart Street (photograph: McDonald Institute Cambridge)

Figure 120 - Ink writing on ballista *balls*
Photographs of 2H and 2E: McDonald Institute Cambridge. Photographs of 3E and 2F by the author, by kind permission of the Sudan and Nubia Department of the British Museum

inscriptions to mark the work of a particular legionary century: OKTAVI 'century of Octavius'. The spelling OKTAVI confirms that the name is in Greek script, as with centurion Pompeius on Ball 2E. **(bottom left)** Ball 2E. Diameter ± 15cm. Line 1: ⋀ I (Greek *iota*) 'weight 10 [*librae*]'. Line 2: P X ' of ten [*librae*] weight' Line 3: ⅂ ΠΟΜΠΗΙΙ 'century of Pompeius'. **(bottom right)** Ball 3E. Diameter 13.1cm min. 17.3 max.. Line 1: ⋀ IΓ (*iota*=10, *gamma*=3). Line 2: P XIII 'weight 13 [*librae*]'. 13 *librae* = 4.25kg. The actual weight today is 3.9kg. As with Ball 3F, the discrepancy may be explained by the drying out of the porous stone in museum storage. **(top right)** Ball 2F, uninscribed, showing the sedimentary layers of the friable Nubian sandstone, accounting for the fact that 50% of the catapult balls had broken during manufacture. The broken pieces would have been available for use as hand thrown missiles.

In a remarkable *tour de force* Hans Barnard has weighed and analysed the sizes and shapes of the 569 balls, of the original 780+, which are still on the site, at the time of publishing this was the first full and detailed record of any large cache of Roman

Shape	Number	Average diameter	Average weight	Average Index
Rough	153	14.2	2.6	0.88168
Well rounded	27	14.6	3.0	1.01840
Flattened	27	14.7	2.6	0.86991
Irregular	26	14.3	2.3	0.78955
Hemispherical	20	14.8	2.6	0.81170
Cubic	18	13.3	2.4	0.98434
Ovoid	9	15.5	3.1	0.82284
? Complete	3	13.5	2.2	0.82431
Greywacke/dolerite*	2	13.3	2.1	1.11526
All**	569	14.3	3.1	1.01840
Standard deviation		1.86539	1.24145	0.16908

Figure 121 - Hans Barnard's size analysis. The seven British Museum balls
(top) Hans Barnard's painstaking analysis of the surviving balls. **(bottom)** The seven balls stored at the British Museum illustrate the difficulties facing the legionaries in trying to achieve a round shape. (photograph: the author, by kind permission of the Sudan and Nubia Department of the British Museum)

Imperial *ballista* balls. Similar large groups of stone ammunition have been found at the Roman sieges of Jerusalem (Arbiv 2023), Gamla (Holley 2014), and Masada (Holley 1994. Stiebel and Magness 2007). Because of the friable nature of the local sedimentary Nubian sandstone (Figure 121) 50% of the balls had broken during manufacture, no doubt causing much fury and frustration to the legionaries during the long hard labour of shaping the stone ammunition. 75% of the pottery found during the excavations is from resin-lined amphorae probably for Egyptian wine, not noted for its quality but no doubt providing the stonecutters with the same reward or palliative as the rations of rum issued by captains in the British navy for uncongenial tasks such as chipping the casting flaws off iron cannon balls.

The ink inscriptions record the differing shot weights used by individual centurions. Every *ballista* was designed by the Hellenistic formula for a shot of a certain weight, and the width of the trough on the slider would have set limits to the stone's maximum diameter. However, it has long been obvious that that a *ballista* might be required to launch stones of lesser weight and diameter than the maximum. Such balls would have to be raised to allow the belt-like bowstring to contact them half way up. The solution proposed by the author is a packing plank to slip onto the base of the slider's

Figure 122 - Clusters of ballista *balls on the West Rampart*
View of the West Rampart cobble deposit, with two clusters of *ballista* balls in the foreground. The gap between the clusters may represent the position of a *ballista*. The rising waters of Lake Nasser allowed the archaeologists' boats to moor alongside
(photograph: McDonald Institute, Cambridge)

channel (Figure 110). Centurion Octavius' name is inscribed on balls of six, eight, and nine *librae*, of diameters 12, 14 and 15cm. A ten *librae* stone, Ball 2E, is also about 15cm. Since there is only a maximum of 3cm size difference, I suggest that Octavius was using a ten *librae* machine, one of the standard sizes on Vitruvius' list, with a packing plank 1.5cm thick to raise the six *librae* missile. Launching lighter missiles than a machine's designed calibre would also extend their range.

If we apply this approach to the 285 unbroken balls listed by Barnard, plus the separately recorded 43 and the seven in the British Museum, 81% could have been launched by 20 and ten *librae* machines, 17% would be suited to a smaller six *librae*

Figure 123 - The Kandaxe ball and Message for Hitler on a WW2 shell

The Kandaxe ball in Figure 123 **(left)** is Ball 4F. Diameter minimum 11.5cm maximum 16.6cm. This weighs 2.8kg, and was possibly a nine librae missile. Line 1: Λ Z. I interpret the many occurrences of Λ (lambda) followed by a second letter such as Γ (gamma), E (epsilon), Z (zeta), Η (eta), as standing for λόχος (lochos) the Greek equivalent of centuria (century), followed by a letter numeral. ΛΟΧΟΣ Z means Number Seven Century. Line 2: ΚΑΝΔΑΞΗ (Kandaxe). Strabo (17.1.54) describes the ruler of the Ethiopians in his time as Queen Kandake. He, and from the evidence of this ball the Roman garrison, took the Meroitic royal title kdke as the personal name of the monarch. The penultimate letter on the ball is xi, not kappa, possibly the way the name was pronounced by the legionaries. Line 3: I ? A ? O N. Surface damage has removed two of the six letters. Only one word fits in: ἱκανόν (hikanon) the neuter of the adjective ἱκανός meaning 'befitting', 'sufficient' etc.. With things it can mean 'large enough'. Here the unexpressed noun it describes is probably βέλος (belos) meaning missile, or δῶρον (doron) meaning gift, present. So lines 2 and 3 are a personal message to the enemy Queen, with her name in the Vocative Case: 'JUST RIGHT FOR YOU, KANDAXE!' This makes it by far the most interesting Roman ballista ball found to date, because its wording reveals the attitude of mind of the artillerymen, encouraged to focus their hatred on the enemy leader. It has immediate parallels with the graffiti on bombs and shells of both World Wars. Line 4: M·V are apparently the initials of the senior centurion, perhaps Marcus Vipsanius (i.e. Agrippa) who has imposed his initials on 13 of the 38 missiles. **(right photo)** Gunner loading a shell with a message for Hitler. Second World War, North African campaign (photograph by kind permission of the Imperial War Museum (IWM negative E20258).

Figure 124 - Qasr Ibrim tanged boltheads.

stone-thrower, and the three 25cm diameter stones and nine of the additional 43 balls would require a 25 *librae* catapult.

In Figure 124 are more tanged bolt-shooter missile heads from Qasr Ibrim of the type in Figure 50 analysed by James and Taylor (one pound coin for scale). Three of them have the four-sided bodkin points, the tang of one has bent on impact; one (left) has a triangular point (compare with the Alésia one in Figure 18), and at bottom right is an example of a blunt point for stunning game or possibly for practising target shooting without using a standard point which might bend. The condition of the objects is amazing, with file marks and blued finish clearly visible, and the fragment of wood shaft and silk whipping surviving on the blunt head. (photograph: the author, by kind permission of the McDonald Institute, Cambridge)

The Qasr Ibrim balls are for small to medium calibre catapults on Vitruvius' list; the heavy stones of up to 80 *librae* (26.2 kg) were only usually used for battering cities like Jerusalem. In any case the height above the surrounding terrain and the Nile would have increased the range and the force of impact of both the stones and the catapult bolts. The friable nature of the sandstone balls, and the fact there were no soft surfaces in the surrounding area to cushion their landing, would mean that they would have ricocheted and fragmented into deadly shrapnel.

In fact they were never launched: when in 22/1 BC the Meroitic queen, '*a masculine sort of a woman, blind in one eye*' (Strabo), approached the fortress with an army of many thousands, she found that Petronius had already arrived and reinforced the defences. Her wise decision to negotiate rather than attack was no doubt based on a shrewd assessment of the superiority of Roman arms and armour, and in particular the destructive power of the artillery. Like the Britons and the Alani her warriors did not wear body armour.

Chapter 11

Artillery in action in the field: Arrian's battle plan

A single detailed Roman battle plan survives to show in detail how artillery was used in action in the field. In AD 134 Flavius Arrianus, the governor of Cappadocia (Eastern Turkey), faced the approach of a large invading army of Sarmatian mounted spearmen, the Scythian Alans, highly respected by Romans and others as skilled riders and fearsome warriors. They were armed with long lances (Latin *conti*, kontoɪ in Arrian) and swords, and many of their horses were protected by scale armour. Tacitus' verdict (*Histories* I.79.2-3) was that, '*No one is as cowardly as them when it comes to fighting on foot, but when they charge in cavalry formations hardly any battle line can resist them.*'

Arrian chose to meet them in a valley, with his Fifteenth and Twelfth Legions and several auxiliary regiments of cavalry, horse-mounted archers, infantry and spearmen. The site of the battle is uncertain, but may have been the pass east of Erzurum on the ancient military road to Armenia (Bosworth 1977, 234; see Google Earth). He positioned the legions in the centre on the valley floor with the higher ground on either wing held by auxiliary infantry with cavalry protection. '*The catapults are to take up positions on both wings, so that they can open fire on the enemy at extreme range*' (Arrian, *Ektaxis* 19). The higher ground will have increased the range of these catapults; some of the bolt-shooters there may have been the easily manoeuvred *cheiroballistrai / manuballistae*. '*The catapults are also to be stationed behind the whole line of battle.*' To be able to fire over the single line of foot archers in front of them, the eight ranks of the legions, and a screen of cavalry, these would have been the larger, more powerful machines, and would surely have included the cart-mounted *carroballistae* seen on Trajan's Column (Figure 59) of which there were 55 in each legion according to Vegetius. However, there may also have been many of the old wooden-frame bolt-shooters still in service, because it is unlikely that all the older machines were withdrawn when the metal-frame catapults were introduced. The stone-throwers would be of the smaller calibres, and either of the traditional Vitruvian wooden-frame design described above or the more powerful Hatra type described in Chapter 14.

'*There must be silence until the enemy come within missile range. Just as they do so, the whole army is to shout in unison the loudest and most blood-curdling battle cry to Enualios [Mars], the catapults are to fire both bolts and stones, the archers arrows, the spearmen are to launch their missiles... Let stones rain down on the enemy from the allied troops on the higher ground. The combined bombardment from all sides is to be as heavy as possible so as to panic the enemy horses and destroy their riders. It is expected that the indescribable volume of missiles will stop the charging Scythians from coming too close to our infantry line.'*

So, a full artillery barrage could include all these other missiles: stones launched from throwing sticks or by hand, slingshot, javelins and spears, and arrows from the bows of foot or mounted archers.

The Alans were faced with riding into a valley of death. Struck first by the wall of sound from the Roman battle-cry, when they were about 400m from the Roman battleline they would already be entering the edge of a hailstorm of heavy bolts, joined moments later by the lighter bolts of the *manuballistae*. From about 200m ricocheting stone catapult balls would be striking them and their horses and the storm would be swollen by the archers' war arrows, by the cross-fire of stones, spears and catapult missiles from the flanks, and eventually by hundreds of javelins. If the Alans ever reached the bristling *pila* of the legionary line, with their massed formation riding at 25-30 kilometres per hour, as Marcus Junkleman suggests (pers. comm.), it would have taken them well over a minute – to use modern timing – to penetrate this *'indescribable volume of missiles'*. Even during this short time every catapult would have shot at least four missiles, and the two legions' artillery could have launched about 600 bolts and 100 bouncing stones.

The only surviving account of what actually happened is the brief entry in the epitome of Dio Cassius (69.15.1) stating that the Alans feared Arrian, the governor of Cappadocia, and halted their campaign; this suggests that the battle may not have taken place. Arrian's tactical dispositions in the valley, with cavalry and artillery reinforcing the auxiliary infantry who were manning the heights, would have made it very difficult for the Alans to outflank the Roman position, forcing them to enter the dense, lethal hailstorm of missiles on the valley floor, a valley of death. The famous cataphract horse scale armour from Dura Europos, exhibited in the British Museum's 2024 exhibition *Legion: Life in the Roman Army*, would have offered little or no protection. We tested scale armour some time ago and the bolts sliced through the scales, sliding up underneath them and through the leather backing. To destroy the horses, even if they wore this armour, is probably why Arrian lined the sides of the valley with artillery and spearmen. Arrian's orders for the legions are for the front rank to hold their *pila* in the launch position so that if the enemy come close *'they can place the metal points of their pila straight into the prime target – the horses' chests'*. The second to fourth ranks are to aim their *pila* wherever targets appear, *'so that they wound horses, kill their riders...when their pila penetrate shields or cataphract armour'*. When their horses were killed, dismounted Alani, as the above Tacitus quote suggests, would have offered no serious resistance.

Chapter 12

Burnswark Roman camps and native hillfort, Dumfriesshire

Figure 125 - Burnswark Hill from the SE, from close to the Eaglesfield crossing of the M74 (photograph: the author)

Figure 126 - Burnswark hillfort and the Roman South Camp (Reproduced with the permission of Cambridge University Collection of Aerial Photography © Copyright reserved.)

Figure 126 is a May 1950 photograph of Burnswark native hillfort, Dumfriesshire, and the Roman South Camp, looking NW. It was taken by the doyen of post-war aerial photography, Kenneth St Joseph, and shows the prominent rampart and ditches of the Roman camp, with the three mounds immediately outside the camp entrances at the foot of the hill. Standard straight traverses outside the nearest and right-hand gates are also visible. The Latin for these protective banks should be *loricae* (Richmond and McIntyre 1934, 52 footnote); a *titulum* is clearly defined as a ditch in *De Munitionibus Castrorum* 49. The square enclosure in the north corner of the camp is now believed to date to the Iron Age. The lines of the collapsed remains of the hillfort's defences, marked on George Jobey's plan in the next figure, are just discernible running to the left of the top of the wood. Jobey's excavations showed that the hillfort's defences were already in ruins when they became the target of Roman missiles. The site's second use was probably as a Roman training camp for practising missile launching and rapid assault (see text below). The three mounds are to be interpreted as artillery platforms; as such their position outside the camp is unique and against Roman army practice.

> *'The Romans do not wait for war to break out before taking up arms. They do not sit around doing nothing in peace time and only get their hands moving when it is necessary. On the contrary, as though they were born with weapons as part of themselves they never take a break from training, and never wait for emergencies to arise. Their training exercises are no less strenuous than real battles. Every day each soldier puts all his energy into training as though he was at war. …It would not be wrong to say that their training exercises are bloodless battles and their battles are bloody training exercises.'*
>
> Josephus, *War against the Jews* III, 72-75.

The last well-known sentence defines the problem: the surviving archaeological evidence from a real siege against a hostile army and a training exercise for the same scenario will probably look identical at first glance. For coverage of the many discussions and views about the site see Breeze 2011,166-180, Jones 2011, 154-156 and Jones 2012, 20-24. An interim report of Reid and Nicholson's Burnswark Project and Excavations is in the *Journal of Roman Archaeology* 32 (2019) 459-477. A Smithsonian Institute and Channel 5 TV programme has been launched on both sides of the Atlantic in autumn 2021 fronted by John Reid, promoting his personal interpretation of the site as a massacre of Britons by the Roman army. The title is *'Massacre on Hadrian's Wall'*. My review of this highly misleading film is in Appendix B.

The positions of the two Roman camps in relation to the native hillfort at Burnswark, (Figures 127 and 128), straddling and menacing it, make it almost certain that they were of siege camp design. Were they constructed in anger to trap a hostile army on the summit, or were they a training exercise in the construction of siege camps? The second problem is deciding whether the Roman missiles found on Burnswark hillfort were part of an assault on a hostile group of Britons or a training exercise/exercises to practise missile launching and manoeuvres such as uphill assault.

> *'Camps must always be built in secure positions, particularly when the enemy are close by. … Care should be taken to avoid having a mountain or higher hill nearby which would give the enemy an advantage if they captured it.'*
>
> Vegetius, *On the Roman Army* I, 22.

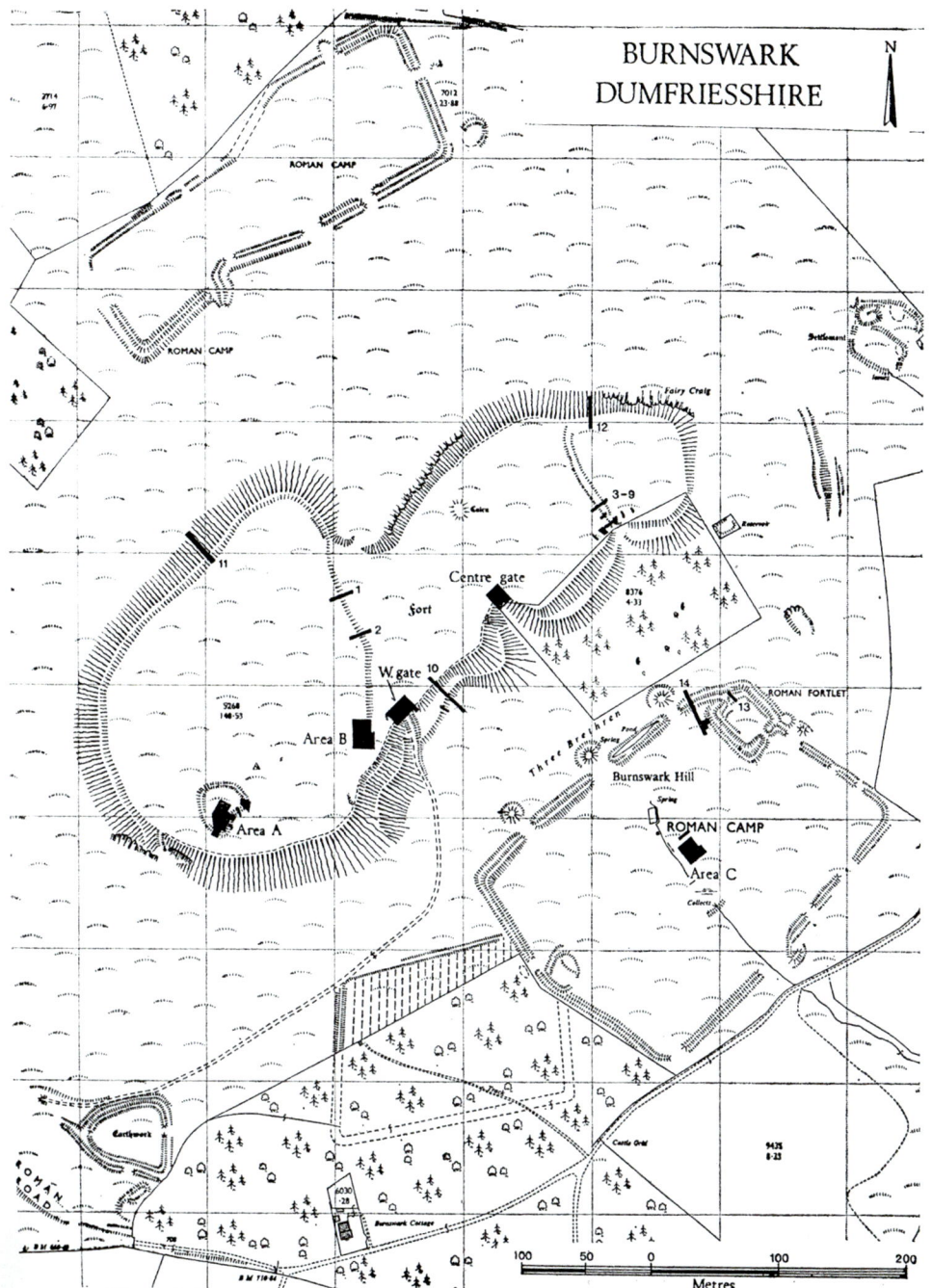

Figure 127 - Burnswark hillfort and Roman camps, Dumfries and Galloway (from Jobey 1977-8 Figure 1)

Figure 128 - The Roman camps and hillfort from the north east (photograph: © Crown Copyright: HES)

The very close proximity of the South Camp to the hillfort rightly led Duncan Campbell, a leading authority on Greek and Roman siegecraft, to argue (2003, 29) that the camp would be at risk of 'rolling stones' attack and that the three mounds outside the Camp's gateways were there to block such missiles. He records the numerous rolling missile attacks on siegeworks from Caesar onwards, and concludes that Roman generals were well aware of the risks of being too near an enemy stronghold. Apollodorus of Damascus, Trajan and Hadrian's famous engineer, gives specific advice in his *Poliorketika* (Wescher 139) on ways to block rolling missiles from such strongholds. The three mounds would not have provided enough coverage to block all such missiles. To make the camp less vulnerable they would have to be high traverses wider than the gateway. See page 166 for further discussion of them.

The aerial photograph above (Figure 128), to which David Breeze drew my attention, gives a much clearer idea than the flat site plan of the South Camp's position immediately underneath the hillfort. It looks down on the Roman camps straddling the hilltop, and reveals the close proximity of the South Camp to the hillfort's rampart, and the exposed positions of the three mounds interpreted as artillery emplacements. The rampart is running next to the right edge of the rectangle of trees; it shows up clearly in Figure 126. Of course, the decision to construct the South Camp in this 'suicidal' position was that of the original commander. Whoever made this decision must have assessed the risk of a 'rolling stones' attack as non-existent.

That the South Camp is not a camp positioned for an actual siege is readily proved by examining the plans of the many Roman siegeworks from Spain to Israel. At none of them is an encircling siege camp anywhere near this close 130-145m proximity to the enemy position.

No actual siege camps have been found in Britain.

The earthworks at Woden Law, once believed to be Roman training siege works, are now identified as Iron Age; and the Roman camps at Cawthorn, formerly identified as practice siege camps, are now identified as two forts and a camp (Halliday 1982, 80-83; Welfare and Swan 1995, 137-42). There is no evidence of siege camps being required to capture any British *oppida*, even those protected by upstanding ramparts and massive ditches as at Maiden Castle. There are none at sites north of Burnswark, such as Traprain Law. Jobey made this point over forty years ago: '*At no time during the course of Roman campaigns in Britain do such elaborate siege-works appear to have been necessary for the capture of native strongholds.*' Jobey 1978, 98. In his day the Woden Law earthworks and Cawthorn camps were still believed to be siege camps; I think today he might cut out the words 'such elaborate'.

Vespasian used barrages from bolt-shooting catapults and direct assault to achieve the capture of more than 20 hillforts of the Durotriges and Dumnonii. I have been unable to locate any catapult balls in these *oppida*. In contrast with the multiple firepower employed in his siege operations in Judaea, in Britain he appears to have relied solely on bolt-shooters to cover his assaults. At Burnswark any mass of unarmoured native resistance fighters on the hilltop could have been massacred in a very short space of time by standard dense salvoes from legionary bolt-shooters, without direct assault. The velocity, weight, accuracy and extreme ranges of the bolts easily outranged and outclassed Iron Age weaponry. They could be launched from positions well beyond the range of enemy missiles. Such a brief but devastating blitzkrieg would not require the construction of siege camps.

Evidence for the abandonment of British hillforts

There is growing archaeological evidence that Iron Age hillforts in Britain were abandoned before the Romans arrived. Breeze's 2019 paper '*Oppida in Britain in the face of the Roman conquest*' establishes (125) that '...we have no evidence for siege camps at these sites [Maiden Castle and Hod Hill] nor at any of the other hillforts or settlements taken by the Romans ... (2019, 125)'. On page 127 he quotes the conclusion by Armit and Ralston 1997, 182 that '*the occupation of hillforts does not appear to characterise the later Iron Age in northern Britain.*'

For Burnswark Breeze concludes (2019, 127) '*...the excavations of George Jobey demonstrated that the hillfort had been abandoned by the time the Roman army arrived. It* *is straining credulity to believe that the Romans besieged an abandoned hillfort; it strains credulity even more to believe that unprotected rebels stood on the hilltop – for what purpose it is hard to imagine.'*

Clear archaeological proof that there is a time gap between the collapse/slighting of the stone rampart face and the artillery bombardment was obtained in 2015 in Trench 2, where a *ballista* ball was found to sit on several centimetres of soil that overlay the collapsed stone work. In the following series of photographs (Figure 129) the centre photo shows the ball as first exposed, embedded in the soil that had accumulated over the stones of the collapsed Iron Age rampart. It has been removed in my right hand photo, and the depression left by the ball is just to the right of the marker stick. The excavator David Devereux has left a cube of the several centimetres of soil between the ball and the stonework, neatly proving that some time had elapsed between the rampart collapse and the missile strike.

The sandstone ball in Figure 129 weighs 0.55kg and displays a flat side, the frequent characteristic of a *ballista* ball. It is within the weight range of missiles found previously at the site, and is close to the size matching the two *librae ballista*, the smallest stone-thrower in Vitruvius' list. It does not sit comfortably in the hand, and is less likely to have been intended for throwing either by hand or by means of a *fustibalus* (arm-extending stick with a sling). The smaller stones of about 0.17 and 0.34kg from Burnswark were probably thrown by hand, or with the *fustibalus*. The centre photo shows the ball as first exposed, embedded in the soil that had accumulated over the stones of the collapsed Iron Age rampart. It has been removed in the right hand photo, and the depression left by the ball is just to the right of the marker stick.

Figure 129. Ballista *ball from the 2015 excavation Trench 2* (photographs: Andrew Nicholson and the author). The evidence of the lapse of time between the collapse/slighting of the rampart and the missile strike.

The dating of the site's features and evidence from the Nicholson and Reid excavations

The small rectangular enclosure in the north corner (Figure 126) has been traditionally identified as an early Antonine fortlet (RCAHMS 1997, 181-2 and Maxwell 1998, 46-9). Because the South Camp overlies this earthwork the Roman camps have until now been dated to the Antonine or post-Antonine period. The enclosure is now believed to be of Iron Age date (Jones 2011, 155) and to have been incorporated into the Roman camp, along with the native circular enclosure which appears to have been responsible for the outward kink in the Camp's south west side. This would mean that the camps could date to any period after the Roman arrival in this area.

The writer of RCAHMS 1997, 182 records *'the clear impression'* that the more substantial uphill sides of the South Camp *'are the result of local remodelling of the defences.'* The following year in his *A Gathering of Eagles* Gordon Maxwell mentions (1998, 47), '... *ballista-firing exercises took place. The heavier weapons occupied massive emplacements at the gateways of a possibly re-used temporary camp...*' His caption to his aerial photograph on p 49 reads, '*The deep shadows of a late summer evening pick out the structural complexities of the artillery training camp at Burnswark, notably the reconstructed north-west side with its three massive ballista-platforms...*'.

One possible interpretation of Jobey's drawing and description of his Cutting 14 across the rampart (1978, Figure 11 and p 81) is that he himself had in fact cut through two phases of rampart. He draws a sharp, distinct difference between his two layers that comprise the rampart, which he calls *'tip lines'*, with the upper one containing fragmented rock, presumably from the ditch cutting into rock. Jobey's photographs would be very important; they may still be available. Excavation of other sides of the South Camp's rampart would provide valuable comparative evidence.

Excavation Directors Andrew Nicholson and John Reid's 2016 Trench 4, excavated by David Devereux, cut into the back of the South Camp's north west rampart. Their published photograph of the side of the section through the rampart (Figure 130) does show a distinct division between two layers forming the rampart; this can be interpreted as support for the fact that it had been heightened at some period, as Maxwell and others had already thought likely from the appearance of this rampart when compared with the state of those on the other three sides. In fact, the ditch on this side is also deeper, as is clear from the aerial photograph (Figure 131 below). Barbour records (1899, 233) that the bottom two feet have been cut into solid rock. On the bottom of the Figure 131 photograph the end of this apparent recutting of the ditch can be seen extending slightly beyond the corner of the Iron Age earthwork.

It is significant that in Figure 130 there is no line of soil deposit between the two distinct layers, in contrast to the depth of soil in Figure 129 above marking the time

Figure 130 - The 2016 Trench 4 cut into the back of the South Camp's north west rampart (photograph: Andrew Nicholson)

lapse after the rampart collapse. This suggests that there was no great interval between the original construction of the camp and the modifications to the rampart and ditch. This lack of visible deposit between the two layers could well account for Jobey's verdict that they were '*tip lines*'.

By finding that lead sling bullets (Figure 135 left) had spilled back from the heightened rampart, the 2016 excavation usefully discovered the actual launch position of the slingers as well as confirming that this missile bombardment took place during the second phase of the site.

Furthermore, the odd clavicular design at the western of the North Camp's entrances (Figure 127) can be argued as a feature only likely to be trialled on a training exercise.

Whoever ordered the construction of the two camps has sited the South Camp as close as the slope up to the hillfort permitted (the slope is shown clearly in Figure 128). This adds further strength to the case for Burnswark being a training area for siege warfare, which the presence of the paving in the South Camp and the two phases in the west gate of the North Camp also support (details in section 7 below). There is no feature of the terrain that would have prevented the camp being sited much further away. The continued gentle slope of the ground to the south east (Figure 131) would even have made it possible to construct the same size camp with its north west rampart on the existing line of the south east rampart. This would have been about the appropriate distancing for an actual siege. Bolt-shooters on this rampart would have had the hillfort top well within range. A camp in this position would have been beyond the range of enemy missiles.

So, there was an unrestricted choice available for the position of the South Camp. I believe that the siting in this very close relation to the hillfort was related to and decided by the effective ranges of slingshot, arrow and *ballista* ball.

The Figure 131 oblique 1990 photograph shows the square Iron Age enclosure incorporated into the north corner of the Roman camp and the circular feature, apparently Iron Age, causing the kink in the line of the south west rampart towards the top of the photograph. The ditch of the rampart at the foot of the hill behind the Three Brethren is visibly deeper and the rampart higher than on the other three sides. The end of the apparent recutting of this ditch can be seen extending slightly beyond the corner of the Iron Age earthwork. Also visible are the standard straight traverses outside the nearest and left-hand gateways. Burnswark Cottage, whose occupier found many lead bullets in her garden, is in the distance.

> ***About a third or a quarter of the recruits ... must always be trained to shoot ... with wooden bows and practice arrows ... Expert instructors must be chosen for this***

Figure 131 - The southern Roman camp, Burnswark, looking south (photograph: © Crown Copyright: HES)

subject ...This skill must be learnt thoroughly and maintained by daily practice and coaching.
The recruits should be given thorough instruction in throwing stones by hand or with slings ... every single recruit must learn this skill by constant practice. After all it is no trouble to carry a sling...'

Vegetius I. 15-16

The sling, bow and the stone-throwing catapult had known <u>effective</u> ranges, all are very similar: Vegetius (II.23) records practice targets for slingers and archers at 600 Roman feet (177m). Importantly, Nicholson and Reid's experiments in tracking *'the most distant shot'* (2019, 470) found that *'...bullets that could be recovered reached 150-170m.'* Philon (*Poliorketika* 84) gives the range (probably the first bounce range) of the

Figure 132 - Burnswark sling bullets, arrowheads and a Trajan's Column stone thrower and slinger

Hellenistic stone-thrower as about 160m (Vitruvius' improved *ballista* design probably added a few metres to that). So, if the South Camp had been sited further south east by, say, 100m, all three types of missiles would have been either out of range of the hillfort or at the limit of their effective range.

The smaller Burnswark stones of about 0.17 and 0.34kg were most likely thrown by hand, or with the arm extension afforded by the *fustibalus*. No ancient writers give the ranges that could be achieved by throwing stones by hand or with the *fustibalus*. The modern record for throwing a cricket ball by hand (0.155kg) is 132.66m.

Burnswark has the types of missile that Vegetius states require constant practice.

In Figure 132 are seven Burnswark lead sling bullets from the Reid and Nicholson excavations. The top two are acorn shapes, the centre three lemon shapes. The bottom two are of a smaller size with a 6mm diameter hole. John Reid has discovered that these make a whirring noise in flight. (photograph: John Reid). At bottom left are some of the arrow heads with triple barbs discovered by George Jobey (Jobey 1978, Fig. 13 and 99-100). On Trajan's Column Scene LXVI Roman auxiliaries are preparing to throw stones by hand and by sling.

The three stone balls and a sling bullet in Figure 133 are from the 2015 excavations. The three balls, all of local sandstone, weigh (left to right) 650g, 340g and 65g respectively. The lemon-shaped sling bullet weighs 54g. The 650g ball is exactly the

Figure 133 - Burnswark missiles (photograph: John Reid)

weight for a two *librae ballista*, the smallest on Vitruvius' list. The 340g stone is small enough to be held in the hand and thrown. The 65g stone is too light and the wrong shape to travel along the slider launch channel of a *ballista*; it may have been intended to be launched by a *fustibalus*, whose sling would perhaps fit neatly around the stone's distinctive middle section.

The missile bombardment

> 'On the available evidence it is impossible to envisage these bullets as having been fired against upstanding hill-fort defences, and no missiles have ever been recorded as coming from beneath the tumble of the ramparts, such as might have been anticipated had they been upstanding when the missiles were fired.'
>
> Jobey 1978, 88.

Jobey's handling of the evidence has stood the test of time, and his view of the missiles as striking the collapsed remains of the hill fort defences has been confirmed by the 2015 excavations.

Excavation Director and experienced slinger Andrew Nicholson has suggested (pers. comm.) that signals in the interior of the hillfort behind the two gateways and elsewhere can be explained as over-shoots, aimed slightly too high for the target of the collapsed rampart. Of the single lead bullet found loose within the hut complex in Area B, Jobey comments (1978, 78) that this *'is hardly sufficient strong evidence that such a house had been a target for assault; other considerations apart, the whole area was sufficiently close to the ramparts and west gateway of the hill-fort to have fallen within the over-zealous or misdirected range of such missiles.'* The difficulty of aiming and following the trajectory of a lead bullet is established by Reid and Nicholson's meticulous study of slinging with lead bullets (2019, 470): *'The maximum range of the larger bullets was*

Figure 134 - The Three Brethren and the 2016 trenches behind the rampart

more difficult to establish as they could not be tracked beyond 30m despite receiving a coat of high visibility paint. Lead bullets were impossible to track visually...'.

The fact that the majority of the missiles land consistently all along the collapsed rampart emphasises that this was the intended target. They can hardly be explained as having fallen short of a target of rebellious Britons on the summit, because that range was also within the maximum ranges for Roman slingshot, arrows and catapults from that position. The line of the collapsed/demolished stonework of the hillfort's rampart forms a coronet-like band just below the summit (Figure 126) and provided a clear target visible from many, but not all, parts of the South Camp's rampart.

In Figure 134 the nearer of the two trenches, Trench 4, was eventually extended into the back of the rampart to reveal possible evidence of the heightening mentioned in the text above and shown in Figure 130. Trench 5 beyond that uncovered two groups of sling bullets dropped by slingers on the rampart top. These lead missiles had rolled back down into two hollows. In Figure 135 (left) the excavation is under way of the larger of the two groups of bullets that had rolled back from the South Camp rampart. The dense cache of 375 lead bullets in the North Camp (right) is believed to be the largest discovered in the Roman Empire. It may originally have been in a container of wood or leather.

Barbour (1896, 213-18) had already recorded finds of missiles in the form of stone balls and sling shot. Jobey's excavation report records a scatter of *ballista* balls, a large number of lead sling bullets and some triple-barbed archers' arrow heads clustered on the stone rubble of the collapsed/slighted rampart facing and at the two gateways (Jobey 1977-8, 86-91). Reid and Nicholson's 2015 trenches confirmed and extended Jobey's results, producing remarkable evidence that dense clusters of signals from lead bullets extended beyond the gateways and almost half a kilometre along the

Figure 135 - Groups of lead sling bullets found in Trench 5 of the 2016 excavations and in the North Camp

native rampart above the South Camp (Reid and Nicholson 219, 473). More stone balls were discovered in 2015-16 and, most importantly, the first ever iron catapult bolthead from the site. Although it is badly corroded and has lost a small amount from the rear of its socket, it appears to be of a size matched to the Xanten-Wardt scorpion catapult, or more likely to its powerful successor the arch strut *manuballista*. Both types of bolt-shooter overlapped and were using the same size of bolthead in the second century.

This find confirms that there was a legionary presence at the bombardment, because while there is evidence that some non-legionary units were occasionally allowed to use stone-throwing catapults (Campbell 1986), there is no evidence that they were ever allowed to use bolt-shooters. A commander with legionary troops at his disposal, as found at Burnswark, would not have wasted time, effort and resources on upgrading the existing South Camp; he would have launched dense, continuous salvoes from his bolt-shooters stationed well away from the base of the hill. Any native opposition forces that might have been present would be of a low threat that, if necessary, could be annihilated easily and quickly without recourse to the level of military commitment implied by the archaeological remains.

The sling was not the prime attack weapon of the Roman army. It was of course a vital missile for the catapult-lacking auxiliary regiments, as evinced by the defence of Velsen (Bosman 1995, 99-103).

In Figure 136 from Reid and Nicholson 2019, black dots mark the distribution of lead signals. Most are grouped along the rampart of the South Camp, with a number from the interior. There are none on the Three Brethren. The lead signals behind the South Camp rampart are assumed to be from bullets accidentally dropped and lost by Roman troops during the bombardment. Even modern replica bullets painted bright

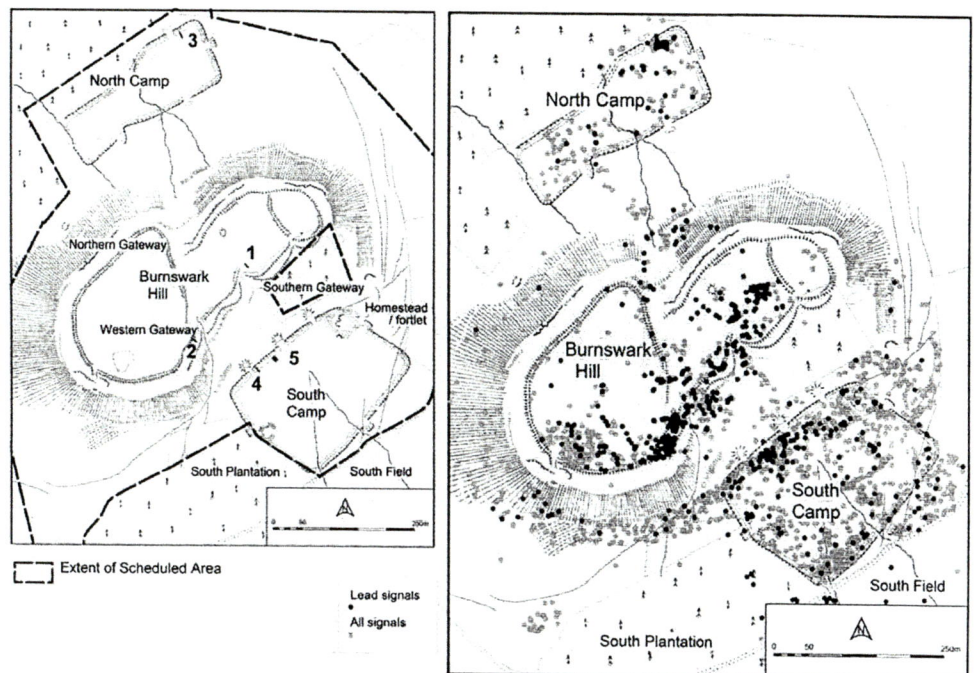

Figure 136 - Excavation trenches 1-5 and the distribution of lead signals from 2013 to 2017

orange are reported as difficult to locate once they have fallen into the grass. On the hillfort a mass of signals clusters along the collapsed/slighted stone rampart facing. Lead signals in the interior of the hillfort behind the two gateways and elsewhere are explained by experienced slinger and Excavation Director Andrew Nicholson as overshoots, aimed slightly too high for the target of the collapsed rampart. In the North Camp the group of signals in the top (north) corner were found in 2016 to include the cache of 375 bullets in Figure 135 above. A scatter of bullets launched from the North Camp seems to have been aimed towards the obvious point of exit from the north west side of the hillfort.

Jobey stopped short of being absolutely certain that the secondary feature of large slabs found by Barbour in the two hill fort gateways had been laid out as targets for missiles by the Romans themselves (Breeze 2011, 170).

Lawrence Keppie (Keppie 2023, 75), in a chapter on the effectiveness of slingers comments, *'The concentration of bullets on the line of the hillfort ramparts at Burnswark, Dumfriesshire, attacked c. AD 142, demonstrates the accuracy of fire from the north rampart of the Roman camp at the foot of the hill, a distance of 150-180 m.'* The diagram (Figure 136 right) does not support this claim of accuracy, with overshoots into the hillfort and a wide variation of range of lead signals. Skilled slingers would have achieved a much tighter grouping of hits along the rampart. I suggest that the spread of hits in

Figure 137 - Carol van Driel-Murray at the centre one of the Three Brethren catapult mounds

Figure 138 - Carol standing in the South Camp's deep recut ditch west of the centre gateway

the diagram resembles the results to be expected from unskilled troops requiring training in slinging.

There are probably many more lead bullets and other missiles to be discovered, but there is no evidence that Roman forces redirected the missile attack and concentrated it beyond the line of the collapsed ramparts onto the top of the hill.

In Figure 137 Carol is standing in the reed-filled semi-circular ditch protecting the uphill side of the mound. The recut ditch and heightened rampart of the South Camp is seen running into the right-hand corner of the photograph. Also visible is the break in the rampart and ditch marking the gateway into the South Camp. The photograph proves that the mounds were positioned immediately in front of the gateways, with the tail of the mound lining up with the outer edge of the Camp's ditch. In Figure 138 the rampart still survives to an impressive height.

The collapse/demolition of the hillfort's rampart

Trench 2 of the 2015 excavations confirmed the results of George Jobey's 1972 excavation, that the missiles had landed on top of the collapsed Iron Age defensive rampart. In Figure 139 David Devereux is showing Lawrence Keppie the stone facing of the Iron Age rampart tumbling down the hillside. This is being removed in the left half of the trench to investigate possible earlier features. None were found. The hillfort has no ditches, as both Barbour and Jobey's excavations had already proved (Barbour 1998, Plate IV; Jobey 1978, Figure 4). The stick close to Lawrence marks the position of the *ballista* ball shown in Figure 129.

As to why the hillfort ramparts had collapsed, Jobey (1978, 67) concludes '... *it is difficult to escape the conclusion that the front of the rampart had been deliberately felled at*

Figure 139. The 2015 excavations on Burnswark Hill

some stage. Whilst the natural gradients are considerable, the downward and forward thrust would hardly be sufficient in itself to dislodge the large blocks of stone forming the bottom course of this face.' The hillfort's defences may have been dismantled under Roman supervision. Other northern hillforts such as Carrock Fell and Ingleborough also have slighted defences.

Occupation of the hillfort after the collapse/demolition of the ramparts

There may have been some Iron Age houses occupied on the hilltop at the time of the bombardment. Jobey found one in his excavation of Area A, and two in Area B, *'one of them overlaid by at least two if not three superimposed houses of ring-trench construction'*. From the discovery of *'some sherds of Roman pottery of the first half of the 2nd century A.D.... it may be inferred that at least the latest of the round timber-built houses was occupied at this time.'* (1978, 78).

The consequence of the loss of the security of the gated stone-faced rampart would have seriously affected the lives of any families who were still living in the hillfort. There was a daily threat to livestock and people from the numerous wild beasts such as wolves and boars that were rampant at the time and dangerous to confront. All livestock would have been taken up into the comparative safety of the hillfort every evening.

It is significant that the Iron Age enclosure incorporated in the corner of the Roman south camp has a strong bank and ditch protection. Even more striking is the small Iron Age riverside enclosure at Boonies near Bentpath in Westerkirk parish, excavated by Jobey (1974, 119-140), where the sequence of roundhouses is protected by high banking and a hefty drop-down door. Andrew Nicholson (pers. comm.) reports that two Iron Age enclosures very similar to that in the corner of the Roman South Camp have been located near Ward Law hillfort at the entrance to the River Nith. All these enclosures may well represent the pattern of the withdrawal from this and other hillforts once their defences had been levelled or they had been abandoned for other reasons.

Because Jobey only identified one house on the top of Burnswark that can be dated to the period of Roman occupation, estimation of the numbers of houses still occupied when the Roman forces arrived will remain unsubstantiated guesswork, unless extensive future excavations test the whole area of the top of the hill. The cost of these and the time taken would be difficult to justify.

Discoveries of sling bullets outside the South Camp

Scatters of sling bullets: Reid and Nicholson's new diagrams of lead bullet distribution (Figure 136) show a scatter of lead signals well clear of the west of the South Camp. The diagrams do not quite extend to the farm to the east of the hillfort or to Burnswark Cottage along the track to the west, where the retired Italian lady artist who painted the dustbins used to live. The cottage is marked on the plan in Figure 127, and can be seen in the distance in Figure 131. She showed me a handful of lead bullets freshly dug up from her garden on two occasions. Ian Caruana discovered 27 bullets when the trees in the wood surrounding the Cottage were felled. A bullet found 'close to the farm' east of the hillfort (Figure 4 bottom edge) had been purchased by a local doctor whom I persuaded to donate it to Annan Museum. Jobey (1978, 87) mentions a casual find of 2 bullets *'as far distant as 70m to the SW of the SW corner of the Roman south camp ...'*

One of the metal detector volunteers who took part in the recent excavations drew attention (pers. comm.) to many isolated lead finds by detectorists in areas well clear of the South Camp, not all of which were handed in to the excavation HQ. He suggested that they represent troops told to wander off and practise moulding bullets. If so, it would suggest that there was no risk to them from hostile natives.

Evidence suggesting that the camps were intended for repeated use

The upgrading of the Roman rampart and ditch at the foot of the hill would have taken many hours, even by a large team. In a section through this rampart Barbour found stone pitching capping part of its top and north faces (1899, Plate V, Figure 1), and records that the bottom two feet of the ditch have been cut into rock. A similar section by Jobey (1978, 81 and Figure 11) includes the suggestion that Barbour's stone capping was derived from the rock cutting of the ditch.

The existence of paving in the centre of the South Camp was discovered by Barbour (1899, 227-8), and confirmed by Jobey, although Jobey failed to find Barbour's stone walling: *'Although it was undoubtedly an area of good quality paving about 6m broad, presumably running for at least 55m (180') as recorded by Barbour, there was not a vestige of his so-called flanking walls found in the 26m length examined...'* (1978, 82).

Barbour also found (1899, 226) that the roadway into the South Camp's east gateway *'...has been surfaced with pavement of heavy stones (Plate VI. Figure. 6), of which about one-half remain on the ground...' 'Partial clearing of the earth from the west one disclosed a roadway surfaced with pavement.'* (1899, 227). This paving implies heavy traffic through the gateways, and the need to provide hard standing inside the camp for wagons and heavy equipment, such as catapults.

At the North Camp Barbour (1899, 233) noted that at the west gateway *'...there was a surfacing of gravel over a heavy stone pavement, partly wanting, below which was another pavement of stone-work of some kind. The lower part of the traverse ditch has been quarried out of very hard rock...'* This suggests two periods of road reinforcement. Again, cutting the bottom of the ditch out of rock would have been a lengthy process.

All these features are indications that, both in their original construction and in the later rampart upgrading, the camps were not intended for brief occupation, but constructed for repeated use.

The date(s) of the missile launchings

It is going to be very difficult to find evidence to date the missile launchings with certainty. There are other lead sling bullets at Birrens, Housesteads, Ambleside and elsewhere. Provisional results of Reid and Nicholson's Carbon 14 and isotopic analysis of the lead bullets are reported to suggest a date range of *c.*135 to the late 150s and a Derbyshire origin (pers. comm. Andrew Nicholson). If this verdict is accurate, the period when launchings could have taken place could be at any time during the governorships of Iulius Severus (131-4) and Mummius Sisenna (134-8) if under Hadrian, and during several governorships from Lollius Urbicus (138-145) to Iulius Verus (155-8) if under Antoninus Pius. The mid-second century dating is now well-proven.

The Three Brethren as artillery platforms

The three mounds, known locally as 'The Three Brethren', are unique in being unlike any other design of traverse protecting camp entrances. Today they measure up to 15m across and 3.5m high and are protected by a curve of ditch on the uphill side (Figure 137). They are widely agreed to be emplacements for artillery, as Collingwood first suggested (1925-26, 52). Interestingly, the pattern of lead signals shows that no sling bullets were found on the mounds, and so it is unlikely that slingers were stationed on them.

Len Morgan and the author have reconstructed the smallest *ballista* on Vitruvius' list, for two *librae* (0.65 kg) stones, suited to launching the larger balls found on this site. This catapult occupies a footprint of 1.68m long, i.e. back to front, and 1.52m laterally, and could easily fit onto the mounds, with ample room for the team of five or more that we find to be the minimum necessary crew for assembling, operating and moving it (Figure 143 below). The larger stone balls from the site vary in weight from 0.6 to 1.1kg.

Campbell (2012, 5) does not believe the Three Brethren mounds to be catapult emplacements and claims that they are traverses similar to the oval mounds outside the gateways of the Reycross camp on the Stainmore Pass. However, the Reycross mounds are in general positioned 60 feet away from the gateways, exactly as recommended for a *titulum* or *lorica* in this situation by *De Munitionibus Castrorum* 49-50 (Richmond and McIntyre 1934, 52). Figure 137 shows that the Three Brethren are positioned immediately in front of the South Camp gateways.

Figure 140 - Trajan's Column Scene LXVI. An arch strut bolt-shooter on a log emplacement

The position of the Three Brethren immediately in front of the three gateways below the hillfort is the most important visual clue that this is not a setup for a real siege. The mounds are most likely part of the original camp construction. Barbour excavated the centre one, stating that '*the traverse, or tower, ... is built over a rough pavement*' (1899, 227 and Plate VI Figure 7). He then states that he has excavated the east one with similar results. The 'pavement' is probably to be interpreted as a foundation layer for the mounds.

First and foremost, not a single one of the numerous examples of Roman siege camps of any period in any country have artillery mounds immediately outside the camp gateways. **It would be against Roman army regulations and practice**. In a siege or battle situation this would have left the catapults dangerously isolated and exposed to capture by enemy assault.

The capture by the Dacians in AD 86 of the artillery and artillery crews of Legio V Alaudae was a never-forgotten disaster; Trajan himself recovered them some 15 years later along with the legion's standard (Dio lxviii 9, 3 and 5; Wilkins 1995, 8). See Figure

57 for the scene on Trajan's Column of two Dacians on a stockade with a captured arch strut catapult. In the siege of Jerusalem Josephus records the Jews making frequent lightning sallies to destroy the catapults with axes or fire (Josephus, *Bell. Jud.* V, 279 - 280, 286–7 etc.).

The standard catapult position for camps is stated by the *De Munitionibus Castrorum* 58: *meminisse oportet ...tormentis tribunalia extruere circum portas, in coxis in loco turrium*. 'One must remember ... to construct platforms for artillery around* [i.e. on either side of] *the gates, at the corners in the position of towers'*. Josephus III, 80 describes *scorpiones, catapultae and ballistae* as stationed on the camp walls in the spaces between the towers.

On the battlefield the artillery is never positioned in front of the infantry: Arrian, when facing the Alani, records that the catapults are to be stationed behind the single line of foot archers in front of them, the eight ranks of the legions, and a screen of cavalry. Vegetius lists (II.15) the *manuballista* as a catapult operating behind the front line of the *antiqua legio* ('legion of earlier times'), and sometimes positioned in the fifth line along with *carroballistae* (III.14).

The Burnswark mounds must surely have been topped with solid flooring, essential for the *ballistae*: our full-size reconstruction weighs about half a tonne. Some sort of side, front and even overhead protection would be desirable, if only to shield the vital sinew-rope springs from the weather. The only clue to such a structure is the catapult emplacement constructed of logs on Trajan's Column Scene LXVI (Figure 140). Lepper and Frere's suggestion (1988, 106-7) that this is an exploded diagram of a roofed enclosure is very attractive.

Another concern, raised by this unique positioning of the mounds immediately outside the siege camp's gates, is that it would put the artillery crews at risk of being struck from behind at point blank range by 'friendly fire' launched from the South Camp rampart by the slingers, archers and others. Of course, the all-round protection suggested above by Lepper and Frere would obviate this risk. The mounds would also be liable to interrupt the line of sight and trajectory of those launching missiles from the rampart. In fact, the pattern of lead signals seems to show that the slingers on the rampart were positioned slightly to the sides of a direct line to the mounds.

If this was a real siege situation with a formidable threat from hostile rebels, it is puzzling that only the rampart and ditch facing the hillfort had been upgraded. One valid conclusion is that only this part of the camp was required for the particular operations.

The combination of the unauthentic out-of-camp exposed artillery mounds, the wide gateways (Figure 141 and the discussion in the next section below), and the refurbishment of only the hillfort side of the camp defences, suggests that this is not a suitable scenario for an actual siege of or assault on hostile Britons, but is an artificial set of features for practising skills in missile launching, uphill assault and so on.

In the surviving fragments of his speeches to the units of the African army Hadrian talks about '... *a long course of training...*', and gives his critical appraisal of a very wide range of manoeuvres and military exercises that he witnessed, from camp building to weapon training, the need to increase the rate and accuracy of shooting with bows, comments on launching stones from slings by the cavalry and so on (Speidel 2006, the best edition). This is the authentic context in which to envisage the possible scenario of training at Burnswark, with the different skills being practised, missile launching from the rampart by archers, slingers and hand stone throwers. Stone-throwing catapults and bolt-shooters may have had their own sessions; others could be organised for assault troops in rapid uphill attacks. The various skills, once perfected, could have been combined in final sessions, enacting attacks from the two camps on an imagined enemy.

Training in missile attack and in rapid assault under covering fire

> 'On present evidence a more satisfactory explanation of the complex might be seen in the presence of a field-exercise area or, as it were, a Command Battle School, offering practice in camp-construction and set piece operations involving the use of a variety of arms, including slings, bows and field-guns.'
>
> Jobey 1978, 99.

Jobey's measured conclusion continues to convince. The interpretation of the features of the site as for 'set piece operations' is also this writer's personal view. Nicholson has made a further point, which can be adduced to support Jobey's interpretation, with this line up at the centre South Camp gateway (Figure 141). He has spaced 13 volunteers as Roman troops in battle kit would have been. He calculates that 1,000

Figure 141 - 'Troops' passing through the centre gateway, thirteen abreast. (photograph: John Reid)

troops could pass through the gateway in two minutes, to use modern timing, and move swiftly uphill, flowing round the sides of the three mounds. In a real siege or war situation this wide gateway would be a dangerously weak point for the defence of this camp. It would have required an overlapping straight traverse to make it less vulnerable. It makes sense as part of army training exercises in fast uphill assault by massed troops.

Covering fire

Another point that Nicholson and I discussed onsite is how far slingers, archers and artillerymen could safely deliver covering fire for such an assault. Regular practice in accurate delivery of covering fire would be essential to eliminate 'friendly fire' casualties. The Trajan's Column Scene XLI (Figure 59) depicts two cart-mounted *carroballistae* shooting over the heads of advancing troops.

Vegetius (I. xvi) states that all recruits should be given thorough instruction in throwing stones by hand or with a sling. Figure 132 from Trajan's Column shows both techniques used in a battle situation. Andrew calculated that the trajectory of sling bullets would allow them to be launched safely from the camp rampart over the heads of troops assaulting Burnswark Hill up to about three quarters of the climb before such covering fire would have to be stopped.

*Figure 142 - The BBC one talent **ballista** launching the final ball (photograph: Margery Wilkins)*

My own experience with the trajectory of stone-throwing catapults is that they could have continued to shoot from the elevated platforms of the three mounds high over the heads of troops assaulting the two gateways until they were almost at their target. Figure 142 of the huge BBC *ballista*'s third and final launch of a 26.2kg (one talent) missile proves that it can be angled to hurl the missile out in a straight line for quite a surprising distance before the stone starts to curve downwards to drop almost vertically onto a target. This invaluable photograph is the only record of the BBC *ballista*'s trajectory, because the bank of side cameras intended to record it failed to fire. A similar trajectory can be achieved with our quarter-scale model. The *carroballistae* scene on Trajan's Column Scene XLI (Figure 59) shows two arch strut catapults delivering covering fire over the heads of assaulting legionaries.

Interpreting the missile bombardment

It is significant that several types of Roman missiles are found at Burnswark. Roman firepower in the attack was proportional to the difficulties faced, the relative strength and potential threat posed by the enemy, the contours of the terrain and the sophistication and strength of any opposing fortress. Only against formidable rebellions, such as those organised on a massive scale by the militarily well-equipped Jews, did Roman generals use the full range of firepower, such as that deployed by Vespasian against the heights of Jotapata (Josephus, *Jewish War* III, 166-8): '*Vespasian ordered his artillery, numbering a total of 160 machines ...to shoot at the defenders on the wall. In a coordinated barrage the catapults sent long bolts whistling through the air, the stone-throwers shot stones weighing one talent, fire was launched and a mass of arrows. This made it impossible for the Jews to man the wall or even the area behind it that was strafed by the missiles. For a mass of Arabian archers, spearmen and slingers was in action along with the artillery*'.

Burnswark was a low hill whose defensive rampart had been slighted or had collapsed long before the Roman missile strike(s) took place. British fighters are dismissed in Vindolanda tablet II, 164 as *nudi* ('*not wearing body armour.*') It has been suggested (Birley A 2002, 50) that the troops referred to may be those from the local Anavionenses mentioned in Tablet III, 1475. Any local Iron Age weaponry was no match for Roman bolt-shooting catapults, whose range, accuracy and strike-power made the defeat of all the tribes of Britain inevitable.

Vespasian appears to have only used bolt-shooters during his capture of over 20 *oppida* of the Durotriges and Dumnonii. It has been suggested (Richmond 1968, 33) that the discovery of iron boltheads from these catapults peppering the chieftain's huts and enclosure at Hod Hill may have contributed to the surrender of this hillfort (Figure 22 above).

The many types of missiles launched at Burnswark Hill, not justified by the likelihood of any formidable military threat, are characteristic of a Roman Battle School range where shooting practice with all types of missiles would occur. Burnswark has the types of missiles that Vegetius states require constant practice. The plethora of sling bullets matches the requirement for all Roman army recruits to be trained to use the sling.

As the quotation from Josephus at the beginning of this article states, daily practice was a standard feature of the Roman army. The Latin word for army, *exercitus*, is a noun derived from the verb *exercere* meaning '*to train by practice*'. The altars dedicated to *disciplinae Augusti*, one of which (RIB 2092) was found at nearby Birrens Fort, are correctly translated '*to the Emperor's instruction/training*'. Altars to *Disciplina* '*are rare in the Roman world, and appear to be confined to the armies of Britain and Africa...its early appearance in Britain* [at Birrens] *links it with the Hadrianic popularization of the conception upon the coinage, and it was no doubt the insistence upon it which earned for the Roman army of Britain its reputation of being the toughest of the army groups.*' Richmond 1962, 189.

To summarise

There is evidence for two phases of use of the two Roman camps: their original construction and a subsequent event or events when only the north west rampart and ditch of the South Camp was refurbished and the collapsed/slighted stone rampart facing became the targets of missile attacks.

Roman armies never needed siege camps to capture any British *oppida*, even those with upstanding ramparts, deep ditches and far steeper ascents than Burnswark.

A commander with legionary troops at his disposal, as found at Burnswark, would not have wasted time, effort and resources on upgrading the existing South Camp; he would have launched dense salvoes from his bolt-shooters stationed well away from the base of the hill. Any native opposition forces that might have been present would be of a low threat that, if necessary, could be annihilated easily and quickly without recourse to the level of military commitment implied by the archaeological remains.

The bolt-shooting catapult, not the sling, was the prime attack weapon of the Roman army. The sling was of course a vital missile for the catapult-lacking auxiliary regiments, as evinced by the defence of Velsen (Bosman 1995). The legionary bolt-shooters, with double the effective range of the sling and with heavier, long shafted, iron-headed missiles, outranged the firepower of all of Rome's enemies: their '*missile is launched with such force that it reaches not less than twice the range of a shot from a bow...*' Procopius (i.21.17).

Neat proof that some time had elapsed between the collapse/demolition of the stone facing of the rampart and the missile bombardment is provided by the discovery of the catapult ball in 2015 Trench 2, with several centimetres of soil between the ball and the stonework. There is growing archaeological evidence that many Iron Age hillforts in northern Britain were abandoned before the Romans arrived.

The position of the Three Brethren mounds immediately outside the gateways of the South Camp is unique, against Roman army practice and found at none of the many siege camps in the Roman world. Catapults were always stationed on the ramparts of Roman siege camps, as contemporary witnesses and military writers confirm. They were never sited in this dangerous position which would have left them exposed to enemy attack.

The wide gateways of the NW rampart of the South Camp would allow very large numbers of troops to pass quickly through them in order to climb the hill. They would be a dangerously weak point in a real siege situation or indeed during any kind of threat from opposition forces, but would facilitate practising rapid hill assault in training exercises.

The combination of the unauthentic out-of-camp exposed artillery mounds and the refurbishment of only the hillfort side of the South Camp defences suggests that this is not a suitable scenario for a siege of or assault on hostile Britons, but is an artificial set of features for practising skills in missile launching, uphill assault and so on. To put it a different way, a real siege would have been conducted following the standard Roman siege layout and procedures, with artillery platforms on the ramparts of the South Camp and with well-protected gateways.

The 'suicidal' positioning of the South Camp up the steep slope under the hillfort suggests that the officer in charge of its construction believed that the risk from a dreaded 'rolling stones' attack was non-existent. The terrain here afforded an unrestricted choice for the position of the South Camp. I believe that the siting in this very close relation to the hillfort was related to and decided by the limit of the effective ranges of slingshot, arrow and *ballista* ball.

The Reid and Nicholson excavations have confirmed Jobey's discovery that the missiles have struck the already collapsed rampart of the hillfort. Their impressive metal-detecting results show that the lead signals from sling bullets extended for almost a half kilometre along the ruins of the rampart (2019, 473). The scatter of lead signals from just inside the hillfort is explained by Andrew Nicholson (pers. comm.) as overshoots. There is no evidence that Roman forces redirected the missile bombardment and extended it beyond the line of the collapsed ramparts onto the top of the hill. There are numerous scatters of sling bullets and lead signals well clear of the South Camp; it is suggested that they were caused by troops told to go off and practise manufacturing bullets and that they were therefore not under threat from enemy action.

A range of different types of Roman missiles has been found: lead sling bullets, iron heads from archers' arrows, small stones for launching by hand or with the *fustibalus* throwing-stick, stones of the size and shape for delivery by *ballistae*, and one corroded iron bolthead from a scorpion catapult. Even the discovery of one catapult bolthead is sufficient to prove that the event(s) included the presence of legionary artillery.

The occurrence of as many as six different types of lead sling bullets points to the possible use of the site as an experimental practice area with tests to discover the effectiveness and characteristics of the different types. Reid and Nicholson have themselves carried out extensive tests with replica bullets to find the same information.

As the evidence stands, however, the very small numbers of missiles delivered by bows and catapults contrast sharply with the vast numbers of lead sling bullets and lead signals. This is the very opposite of the tally at an actual siege or indeed at any form of Roman assault involving legionary troops, where catapult missiles, particularly from long-range bolt-shooters, would have left behind vast numbers of iron boltheads to be discovered in Barbour, Jobey and Nicholson's trenches, or by field walking, without the need for metal detectors.

It is argued that the present evidence matches Vegetius' requirement to train all troops in the use of the sling, as would be fulfilled at a Roman Battle School practice range. Burnswark has the types of missiles that Vegetius states require constant practice. With standard battlefield positions for archers and artillery as stationed behind several ranks of legionary infantry, as in Arrian's battle plan, regular practice in delivering accurate covering fire would also have been essential to eliminate 'friendly fire' casualties.

What the three series of excavations at Burnswark have uncovered is the story of a pair of practice siege camps, the southern of which received modifications at a subsequent date. The application of technologies not available to Jobey, including sophisticated metal detectors and isotopic analysis of lead, has produced a startling and detailed picture of the collapsed/slighted rampart of an Iron Age hillfort peppered with lead sling bullets and the target of other types of Roman missile, all launched from the modified rampart.

Nicholson and Reid's extensive lead bullet finds, added to the previous tally of lead bullets found here, make this the site with the largest number of lead bullet signals in the Roman Empire. Their detailed study of the lead finds and their experiments with slinging replica bullets are a *locus classicus* for the performance of slingers.

It can also be claimed that Burnswark is extremely important, in line with Jobey's judgement, as a site unique to date in the Roman Empire (Breeze 2011, 179-80) for understanding the Roman army's training procedures at what today would be termed a Command Battle School for missile target practice and rapid uphill assaults.

There may well have been some Iron Age houses still occupied on the hilltop at the time. Jobey found that from the discovery of *'some sherds of Roman pottery of the first half of the 2nd century A.D.... it may be inferred that at least the latest of the round timber-built houses was occupied at this time.'* (1978, 78). However, once the defences had been demolished the loss of daily protection from the constant threat to livestock and families from wild beasts would in itself have made the hilltop unsafe. It is suggested that the Iron Age enclosure incorporated in the Roman South Camp and the small enclosure at Boonies represent the pattern of small well-protected settlements that followed the abandonment of this and other hillforts. The heavy-duty door on the Boonies' protective banking is particularly telling.

'It is straining credulity to believe that the Romans besieged an abandoned hillfort; it strains credulity even more to believe that unprotected rebels stood on the hilltop – for what purpose it is hard to imagine.' (Breeze 2019, 127). This is a point regularly raised during question time at talks on this subject. I too find it unbelievable that unarmoured Britons would

Figure 143 - A two librae *Vitruvian* ballista *being assembled by a team of legionaries (photograph: the Roman Military Research Society)*

have stood exposed on the line of the collapsed rampart, waiting to receive a hail of missiles whose lethal accuracy was only too well-known from many recent conflicts. (See Point 4 in Appendix B in the Review of Reid's TV film).

This revised Burnswark paper owes much to constructive advice from David Breeze and Bill Hanson.

In Figure 143 the Morgan-Wilkins Mark 1 reconstruction of Vitruvius' two *librae ballista*, is being assembled by an *optio* and a five-man crew from the Roman Military Research Society. This scene gives an indication of the amount of space required to operate this size of stone-thrower, about 3.5m by 5m, and confirms that it would fit comfortably on the Burnswark mounds, the centre and east ones of which were excavated by Barbour and found to be '60 feet in diameter at the base' (Christison, Barbour and Anderson 1998, 223 and Plate VI). The main platform at the top of the mounds would, whatever its design, be much smaller. The two *librae* version is the size of machine that matches the larger stone balls found on the site (Figure 133 above). We find that a crew of five is the minimum needed to cope with joining the heavy front frame to the stock and lifting it onto the base. The total weight of catapult and base is about half a tonne, and this is the smallest *ballista* on Vitruvius' list!

High Rochester *ballistaria*, Northumberland

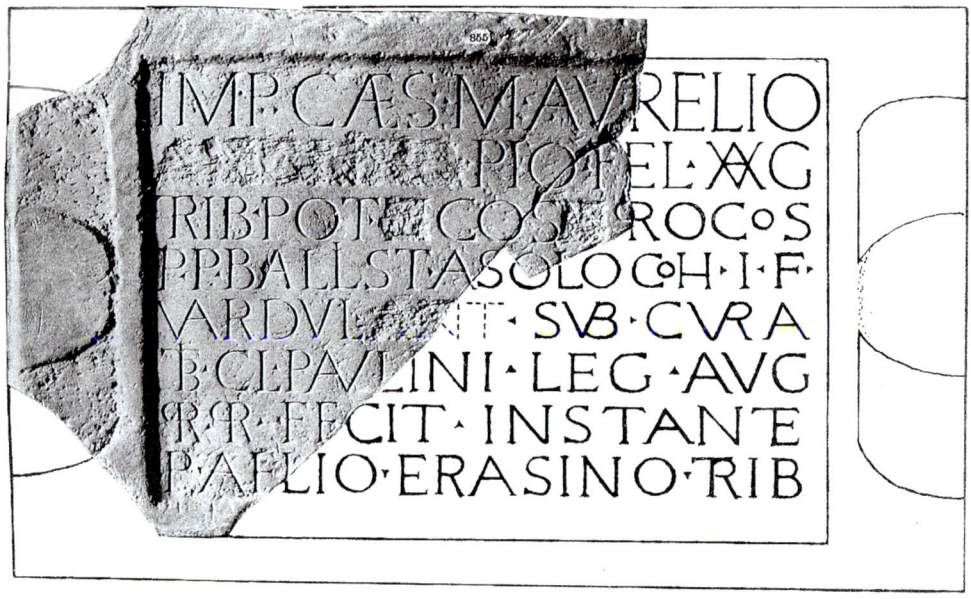

Figure 144. High Rochester ballistaria *inscription (after Richmond).*

Figure 144 is of one of the two inscriptions from the fort at High Rochester, Northumberland, recording the construction of one or more *ballistaria* by the First Loyal Cohort of the Vardulli, a 1,000 strong, part-mounted infantry unit. This inscription, RIB 1280, dates from AD 220, the second, RIB 1281, from 225-35. The design and positions of the *ballistaria* are unknown. Were they simply sheds for catapults, or platforms to raise catapults to a firing position, perhaps with overhead covers and frontal protection? They may have been intended for bolt-shooters, or for stone-throwers of Vitruvian or Hatran type, as attested by the one talent and one-and-a-half talent shot found at this site and at the nearby fort of Risingham. These are on display at The Great North Museum in Newcastle. The presence of one-arm *onagri* suggested by Sir Ian Richmond cannot be ruled out. Many years ago the author was told by a local resident that a 'round grapefruit-sized stone' had been dug up inside the fort; if this is Roman, and not a mediaeval cannon ball, it would imply the presence of small calibre *ballistae*. It seems quite possible that auxiliary units such as the Vardulli were allowed by this date to operate stone-throwing catapults (Campbell 1986).

Chapter 13

The last stone-throwers

The one-arm catapult

The manuscript diagram in Figure 145 (Wescher, 252 Figure XCVII) is of a one-arm stone-thrower mounted on a battering ram suspended from a portable siege tower. Note the three-section ladder construction locked together by long pins, an indication of the structure's portability. Such a ladder-tower could be rapidly set up for mounting an artillery piece in a situation such as that suggested by Sir Ian Richmond for the catapult attack on Hod Hill (Chapter 4)

The most elusive Graeco-Roman catapult is the Greek *monangkon* ('one-arm'), known in late Roman times as the *onager* ('wild ass'). It is first mentioned by Philon (*Poliorketika*, 91) in a passage on defending a city under siege: '*You must try to smash through the roofs*

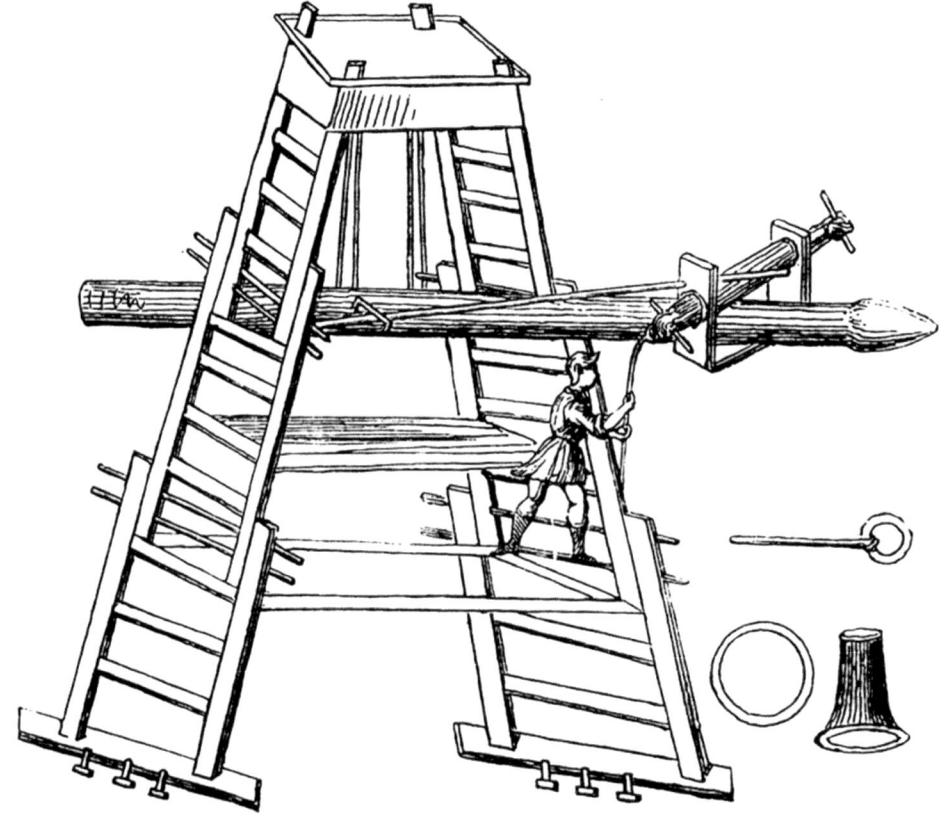

Figure 145 - Ms diagram of a one-arm stone-thrower on a siege tower

The last stone-throwers

of the siege-sheds by releasing very large stones from machines and projecting beams, and by shooting from above with one-talent and one-arm stone-throwers...' This defines the one-arm as a heavy stone-thrower on a par with the very powerful one talent catapult.

There is an intriguing comment by Vitruvius in the introduction to his *De Architectura* (I, 2) where he says that along with three others he was '*put in charge of the construction and repair of ballistas, scorpions and the other torsion machines*' (*reliquorumque tormentorum*). What other torsion catapults existed? Today we only know of the *onager*.

Nothing more is heard of it until the early 2nd century AD when Trajan's engineer Apollodorus (Wescher, 188) describes a one-arm mounted on the front of a battering ram which is suspended from siege-ladders. '*When this ram is brought to bear on the wall by swinging it, it will release a trigger and fire the one-arm at the defenders, and will result in the destruction of many of those who stand in its path.*' The text attributed to Anonymus Byzantinus (Wescher, 252) amplifies Apollodorus' description, and includes an illustration (Figure 145). Again, the implication is that the one-arm is capable of releasing an extremely heavy stone.

This is confirmed by Ammianus Marcellinus, the fourth century historian and army officer, who operated these stone-throwers in Emperor Julian's campaign against the Persians (AD 359 – 363). He gives a general description of the machine, with its single vertical wooden pole (arm) inserted into the single skein of rope, and striking a huge buffer of goat's hair cloth padded with fine chaff. A sling of iron or flax hangs from hooks on the top of the arm (Ammianus XXIII, 4. 5). '*When the battle begins, a round stone is placed in the sling and four young men on each side wind back the bars ... and pull the pole down almost horizontal. Finally, the master artilleryman, standing by and rising to his full height, disengages the pin ... with a sledgehammer; ... the pole is released ... and when it contacts the softness of the cloth it hurls the stone, which will smash whatever it hits.*' (Ammianus History XXIII, 4. 6). Unfortunately, Ammianus does not describe how the huge buffer is supported. There are differing modern solutions to this, the most convincing of which is the very first attempt, by General de Reffye (Figure 146 top). The frame supporting de Reffye's sack is made as a separate item. Our version of the *onager*, constructed by Len Morgan, follows de Reffye's design. Figure 147 is of our scale- model. At the time of writing, we have not been able to assemble and test Len's very large version, which has been delivered to Comlongon Castle, Dumfries, its permanent home (Figure 164).

The technical value of Sir Ralph Payne-Gallwey's experiments with a 'railway buffer' design is that they proved the huge advantage of the sling: his slingless two-ton *onager* hurled a 3.64kg stone from 320 to 330m, but with a sling to 412 to 420m, and at full tension to nearly 457m (Payne-Gallwey 1903, Appendix 11). His machine is now in the Royal Armouries at Fort Nelson, Hampshire.

Figure 146 - Two modern reconstructions of the onager (**top**) General de Reffye's impressive interpretation of the onager from Ammianus' description. (**bottom**) A version built by the Ermine Street Guard, based on the 'railway buffer' reconstruction by Sir Ralph Payne-Gallwey. (photograph: The Ermine Street Guard.)

Ammianus' text clearly links the statement '*it is positioned on top of a mound of turves or a pile of bricks*' with the explanation '*for (**nam**) if it is put on a stone wall a mass of this kind smashes whatever it finds beneath it because of its violent recoil and not because of its weight.*' (Ammianus XXIII, 4. 5). This surely rules out Michael Lewis' radically different interpretation (Hart and Lewis 1986) that changes the Latin text and places the mound of turves or thin bricks as a support for the buffer in front of the machine. The one-arm is the only catapult that rocks on release and requires a resilient mounting.

The enormous power of Ammianus' machine is confirmed by the fact that it took eight men to wind back the arm, and a sledgehammer blow to overcome the friction on the trigger. It is quite possible that a very large *onager* could project even heavier stones than a three talent two-arm stone-thrower. A more lurid proof of the *onager*'s power is Ammianus' account of the engineer standing behind one who was killed when an artilleryman was careless in loading the stone in the sling and it shot backwards, destroying the engineer so completely that not all the pieces of his body could be identified (Ammianus XXIV, 4. 28). Ammianus emphasises the *onager*'s lack of mobility in his description of the difficulties of moving four *onagri* overnight to a new defensive position; the effort was worthwhile because the *onagri* smashed the iron-clad Persian siege-towers (Ammianus XIX, 7. 6-7).

In spite of the dearth of references to this catapult in surviving sources, it may well have been used in sieges from the Hellenistic period onwards. It was much simpler to make, adjust and operate than the two-arm machines; it only had one spring and so

Figure 147 - The Morgan-Wilkins onager

may not have used as much sinew-rope. It could probably cope with heavier missiles, and by adjusting the length of its sling and so its trajectory, it could be made to lob missiles over high battlements or bowl them along the ground against troops.

The traction trebuchet

The traction trebuchet (Figure 148) was in use in China from the third century BC. The date of its arrival in the east mediterranean is not certain, but it may have arrived there by the late sixth or early seventh century AD. It was most likely brought over into the Byzantine world by the nomadic Avars.

The design was simple: a long wooden lever pivoting on a high wooden base, with a sling for the stone missile at one end pulled down by a team of men at the other end. In the manuscript illustration the stone in the sling can be seen hanging next to the right-hand black tent. The manpower to operate it was always available. It was simple and cheap to build, and avoided the complexity of the high-maintenance rope springs and the sophisticated metal work of the *onager*.

It ushered in the end of Roman torsion stone-throwing artillery.

Figure 148 - The Traction trebuchet (Wikipedia Commons)

Chapter 14

The Hatra stone-thrower and the inward-swinging arms theory

A few parts of the spring-frame of an unfamiliar design of catapult dating from the third century AD were found at Hatra in Iraq in 1972. Hatra's superb defences and artillery had driven Trajan's army away in AD 117. When Emperor Severus assaulted the city in 198, his men suffered very heavy casualties and his catapults were burned by artillery fire. His second assault failed; the historian Dio records (LXXVI.11) that *'he lost...all his catapults, except those constructed by Priscus, and many troops.'* The Hatra artillery was so effective at long range that *'... they even hit many of Severus' bodyguards, shooting two missiles simultaneously from the same discharge...'* The height of the Hatra city walls would have increased their catapults' range. When Roman troops approached the wall and began to breach it the Hatreni *'amongst other things shot bituminous naphtha at them and burned the artillery and all the soldiers struck by it'*. The naphtha was probably delivered by the special artillery bolts with 'cage' points (Figure 53).

Figure 149 - Excavation photo of the Hatra catapult (photograph: W.I. Al-Salihi, Directorate-General of Antiquities, Baghdad.)

The Hatra frame and its bronze fittings is shown in Figure 149 as excavated. The photograph is vital as the main evidence for what little has actually survived of this catapult. It had fallen down to the rear of the city wall, apparently from Tower XIX, and landed face down, with the case and slider pointing up in the air. Case, slider and all but a few fragments of the hardwood frame have rotted away, leaving only the 2mm thick bronze sheets that clad the front of the frame, some cast bronze caps from the corners of the frame, and three of the four bronze washers with their underplates and badly corroded iron washer bars. One washer has rolled away on impact to the left side of the photograph. A fragment of the wood of the end of the case, emphasised by the strong shadow, is visible in the centre of the lower left frame, where Baatz records (*Britannia* 1978, 6) 'strong iron bolts and two heavy iron bands' that fixed the case to this lower beam of the frame.

The Hatra catapult, the Vitruvian *ballista*'s successor?

This 1972 find belongs to the last phase of the city's life when it was suffering attacks from the Sassanids. The Hatreni appear to have asked for Roman help because two inscriptions record the presence of a Roman cohort, *cohors IX Mauretanorum,* under Gordian III (AD 238 – 244). The catapult may have been one of the Hatreni's own, but is equally likely to have been a Roman army one. One feature, the fact that its springs are mounted far apart in a very wide frame (Figure 149), is a characteristic of the arch strut *cheiroballistra*, and lends support to the theory that this stone-thrower may be a Roman army redesign of the Vitruvian *ballista*, a parallel exercise to their work on the *cheiroballistra*. The extreme first bounce ranges achieved by the stone-thrower(s) at De Meern (Figure 115) could not be achieved by the Vitruvian *ballista*, and are strong evidence that this more powerful Hatra wide-frame *ballista* was its replacement in the Roman army.

Fragments of the wood of the frame have been identified as from the Caucasian Wing-nut tree, a type of walnut. The front face was covered with light bronze sheets 2mm thick. An all-metal spring-frame of this size would have been extremely heavy and impractical. Unfortunately, only this front plating of the spring-frame survives, plus narrow plates on the front of the frame's sides, together with three of its superb bronze washers with underplates (Figure 152) and some cast end-caps (Figure 150). The catapult had fallen, or was pushed, from the rear of the city wall behind Tower XIX, landing face downwards with its case and slider vertical. The latter and the back of the spring-frame have rotted away. Professor Baatz showed me the Al-Sahili photograph and pointed out that the remains of the stump of the wood case is visible on the lower cross-member in the photograph (Figure 149). '*Exactly in the centre of this cross-member ...were found strong iron bolts and two heavy iron bands. In all probability they fixed the case to the frame. If this is right, the machine will have fallen with its front on the ground and its case or stock sticking up in the air.*' (Baatz 1978, 6).

This photograph shows that the 2mm thick front plates have not been distorted on landing. Baatz' vital account (*Britannia* 1978, 3-9) of the finds, and his discussion of the machine should be studied in conjunction with the Al-Sahili excavation photograph.

When attempting to reconstruct this machine it is vital to be clear about how little has survived. Baatz rightly regretted (pers. comm.) that he or another experienced artillery researcher had not been present to watch the excavation of the catapult. He visited the Mogul Museum in December 1975, three years after the excavation, when he was able to take measurements of the metal plating etc. from their position on the Museum's strange reconstruction, seen here in Figure 150. His exemplary, detailed report and the Al-Sahili photograph are vital to clarifying which parts of the Museum reconstruction of the wooden frame are supported by the evidence.

Figure 150: the surviving bronze front and side plates, washers with underplates, and corner caps, as restored in the Mosul Museum, Baghdad. Astonishingly, this ridiculous restoration ignores the washers and their bars and places the plating face upwards, imagining it to be from the frame of a one-arm catapult. Again, it is important to

Figure 150 - Mosul Museum's strange reconstruction of the Hatra catapult (photograph: Mary Desbruslais)

realise that the details of the back (here the bottom) of the frame are guesswork by the excavators/museum officials. The reconstruction appears to be convincing for the front part of the framework, where the plating surely gives at least the thickness of the hardwood beams to which it had been attached, and the width of the side stanchions. However, with nothing of the rest of the frame having survived, the Museum reconstruction of the rear framework, which Baatz has rightly marked in dotted lines in his drawings, is entirely guesswork. It may well be on the right lines in its assumption that the rear of the wood frame was very similar to the front.

It is clear that there is simply insufficient evidence to be certain what the machine looked like. I would like to stress that the only reason why we have built two versions, with inswinging and outswinging arms, is to test the validity of the 'inswinging arms theory', and not to offer a likely interpretation of the Hatra machine's original appearance.

The modern theory that this catapult had inward-swinging arms is based on the width of the frame and the position of the semicircular cut outs on the front plates (Figure 151). However, inward-swinging arms would have required the spring-frame to have been moved from the front to the middle of the case; and with the case projecting from the front of the spring-frame, it would have been quite impossible for the falling catapult to have come to rest, as it has, on the front of its spring-frame. The plates

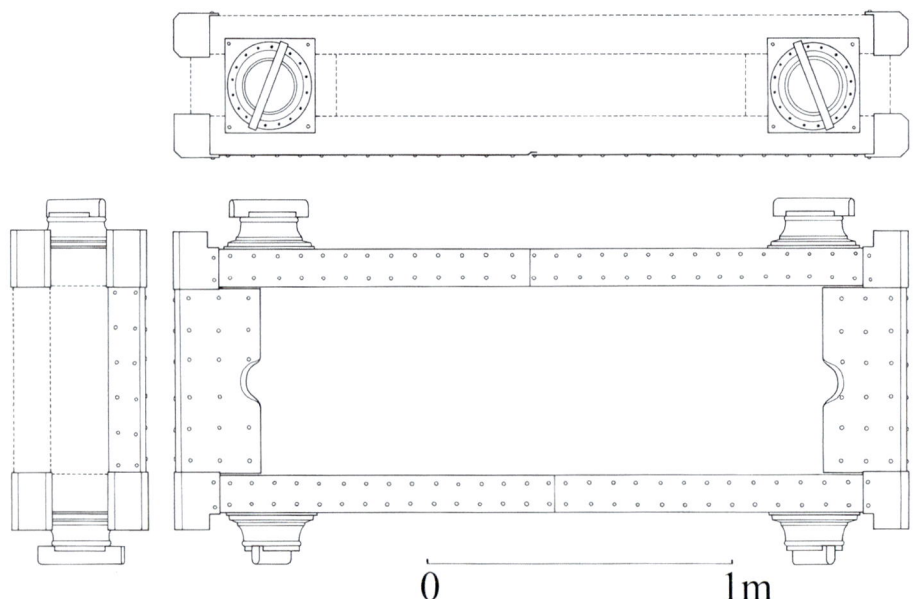

Figure 151 - Baatz's diagram of the Hatra frame
Dietwulf Baatz's drawing of the Hatra spring-frame, based on his measurements of the Mosul Museum reconstruction. He has been careful to use dotted lines to show the possible design of the configuration of the missing woodwork of the frame as suggested by the Museum reconstruction. Certainty about this is not possible (see following text).

on the front of the frames are undistorted. Baatz discussion of this aspect (1978, 6) quoted above should be noted: his reconstructed side view of the catapult is based on the fact that the spring-frame was mounted at the front of the case.

The author believes that the position of the cut-outs on the front upright plates are evidence that the engineers had somehow found a different answer (Figure 152), not necessarily an inward-swinging arm one, to the problem of allowing an increased arc of travel for the arms without weakening the spring-frame. The following Morgan-Wilkins reconstruction uses spring-frames with L-section stanchions in a design which is in effect a sideways stretched version of the Vitruvian *ballista*.

Len Morgan points out that the layout of the spring-frames on our outswinger reconstruction conforms exactly to the slant of Vitruvius' and Heron's rhombus diagrams (Figure 100).

Figure 152 is my suggested interpretation of the Hatra spring-frame, with L-shaped stanchions marked in black. By turning the wood stanchions of the Vitruvian *ballista* (Figure 105) through 90°, frame intrusion on arm travel is markedly reduced, allowing

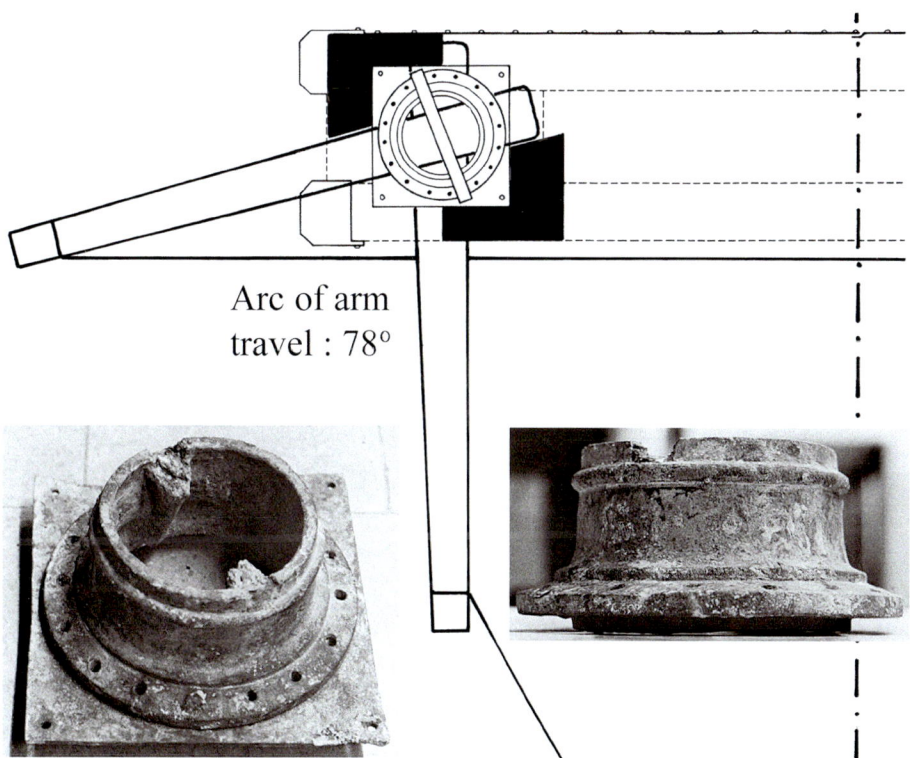

Figure 152 - The author's suggested spring-frame design of the Hatra catapult. The bronze washers. (photographs: Diewulf Baatz)

the bowstring to travel further forward before approaching the rear stanchion. The two spring-frames are 2.25 times further apart than those on the Vitruvian ballista, postponing the moment when the critical 150° final bowstring-to-arm angle is reached and so enabling the arms to be pulled further to the rear. The total gain is a vital extra 19-20° of arm travel, and, as for the *cheiroballistra*, an improved, unrestricted view of the target and so more accurate aiming. Our reconstruction (Figures 152 and 153) is in effect a sideways-stretched version of Vitruvius' *ballista*, with the configuration of the two spring-frames conforming to the *ballista* rhombus diagram of Vitruvius and Heron (Figure 100). The bronze washers are superbly designed and cast, with an outer reinforcing ring below the top, and inner lugs supporting the iron washer bars against the compressive force of the spring-cord. Each washer has a square underplate protecting the wooden frame against wear. The 16cm spring diameter (6.5 Roman inches) is similar to that of a Vitruvian five *librae ballista*, but the volume of the spring-cord will have been less because the ratio of spring diameter to spring height is only 1:6.65, as against the Vitruvian 1:9.4.

In Figure 153 Len Morgan is winding back the arms of his highly effective quarter-scale model. Both Len and Tom's models use weighted golf ball ammunition. The maximum arm arc on this version is about 74-75°, 19-20° more than that of the arms of our Vitruvian *ballista*. The overall length of the bowstring is 91.5 cm, twice that of the inswinger at 46cm. The bowstring travels 57.5 cm along the slider, much further than the 40cm travel of the inswinger's looped bowstring. The length of the arms is 31.5cm, the inswinger's are 23cm long. This means that the inswinger arms are at a massive disadvantage being short levers attempting to wind an extra 30 degrees onto the rope-springs.

Of course, the total absence of the details of the rear of the frame and case makes it impossible to reconstruct it with any certainty. This is one reason why I said (Wilkins 2000, 100) that John Anstee's premise that the Hatra stone-thrower <u>must</u> have had inward swinging arms was erroneous.

Figure 153 - The Morgan-Wilkins outward swinging Hatra catapult

The Hatra stone-thrower and the inward-swinging arms theory

For this 2.4m wide frame the thickness of the missing hardwood cross beams, estimated from the evidence of the plating, is about 12cm (five Roman inches). There may well, as Baatz suggests (1978, 4) have been other transverse beams on the frame. Len Morgan has also added an intermediate set of uprights within the frame (Figure 153), plus a pair of long bracing struts linked to the rear of the case, as found on Vitruvius' *ballista* (Figure 107). Only hard evidence from a similar future archaeological discovery will provide certain answers.

The inward-swinging arms theory

A frequent modern reaction to seeing the wide spacing of the Hatra and the *cheiroballistra*'s spring frames is to assume that the change in design was to allow the introduction of inward swinging arms, a nineteenth century assumption by Victor Prou (Chapter 6 page 70) for which there is no hard ancient evidence. Some scholars have also searched for evidence that the Roman army's *palintone* stone-throwers, on which the arch strut design change is partly based, had inward-swinging arms. Iriarte's summary in 2003 (2003, 137) was that, '*With the information currently available, it is not possible to affirm categorically whether all...palintones were either outswingers or inswingers.*' Since then my revised edition of Vitruvius' text and Len Morgan's full-size reconstructions (Chapter 7) confirm that Vitruvius' *ballista* had outward swinging arms. Also, Heron's diagram (Figure 102) clearly shows the stone-thrower's outward swinging arms, and he states (*Bel.* 100.2), '*When the Half-springs are armed and the Arms recoil <u>outwards</u>* [*eis to ektos*]...' Forwards would have beeen *eis to prosthen* in Greek.

Further evidence that the arch strut catapults' arms were of the standard outward swinging type can be adduced from the fact that the semi-circular cut-outs for the arms are on the outside of the pair of metal spring frames in the *Cheiroballistra* mss drawings, i.e. in the same position as the cut-outs on all the earlier Vitruvian wooden frames (Figures 31 and 36). Confirmation that these mss drawings are showing this outside position comes from the fact that Heron's *ballista* diagram (Figure 102) draws the cut-outs in precisely the same position.

Furthermore, if inward swinging arms had ever existed they would have required the spring-frame to be positioned about halfway along the case, as in Michael Lewis and Aitor Iriarte's diagrams (Figures 155 and 156). From the distinctive design of its stand, the Cupid Gem (Figure 85) appears to be based on that of the Trajan's Column/*cheiroballistra* type, and not the Vitruvian. The Gem shows the spring-frames mounted right at the front of the case, as on the earlier Ampurias, Caminreal, Cremona and Xanten-Wardt machines and the Vedennius relief.

There is also an important feature on the outside of three of the catapults on Trajan's Column (Chapter 6, Figure 96) which I have suggested was a 'sleeve' for an outswinging arm. Such an outside projecting feature would not be required for an inswinging design as drawn by Lewis and Aitor Iriarte.

In my 2000 article (JRMES 2000, 100) I raised the issue of the totally different geometry of an inswinging design, with its looped bowstring and arm movement forward of the frame. I mocked up inswinging arms and bowstring on my *cheiroballistra* reconstruction, based on the layout in Victor Prou's drawing (Figure 56); the drawings by Lewis and Iriarte had not yet appeared. My four points were:

1. To avoid clashing, the arms have to be reduced by about 25%. Iriarte naughtily, perhaps wilfully, only quoted this point (Iriarte 2003, 117, footnote 23), which he countered, and did not mention my other points. In fact, the difference is slightly greater on our quarter-scale models: outswinger arm 7 x 45mm = 31.5cm; inswinger arm 23cm. The outswinger arm has been reduced by 33% for the inswinging version.

2. The bowstring has to be halved. Our outswinger bowstring is 91.5cm, the inswinging version is 46cm.

3. The bowstring travels along the slider only for about two thirds of the distance that outswinging arms allow. Outswinger bowstring travels 57.5cm; inswinger bowstring travels 40cm.

4. For much of their travel, when they are forward of the field-frames (Figure 158), inswinging arms are moving outwards more than forwards, slowing down the forward movement of the looped bowstring, and possibly causing the bolt/stone to leave the bowstring before the latter has completed its travel. This point could be confirmed by high-speed photography (hiring a HS camera proved far too expensive for our tests) which has already confirmed (Wild Dreams TV programme 'Ancient Ballistics') that the bolt in an outswinging bolt-shooter only leaves the bowstring when the arms have completed their travel.

Long before our two large models of the Hatra catapult were completed, an independent review was undertaken of the different performances of the bowstrings by physicist Dr Martin McAree. His graphs (Figure 154) show an alarming drop in the forward progress and velocity of the looped bowstring of the inswinger during the 30 to 80 degrees of arm travel. My caveat 4 above seems justified.

Two further caveats:

5. Any inswinging design would have to provide the same system of applying the brakes to the arms at the end of their travel, i.e. the inner ends of the arms must strike some sort of heelpad on the stanchion of the field-frame just before the bowstring reaches its maximum length. The bowstring, if used as the only brake, would be

Figure 154 - Martin McAree's graphs of the relative performances of inswinger and outswinger

subject to rapid wear and possibly early failure. This is why my first version of the Hatra drawing (Wilkins 2003, Figure. 55) was rightly criticised as lacking this heelpad.

6. If, as has been suggested (Lewis and Iriarte), the inswinging design would add a further 30 degrees or more of travel to the existing 70 degrees of the outswinging *cheiroballistra*, the force required to winch back the arms those extra 30 degrees would be very considerable, and put a very great strain on the catapult's frame, arms and bowstring. We were alerted and concerned by the creaking noises from our inswinger from the 70-degree position of the arms onwards. It took a great deal of effort to pull it back close to the 90-degree mark (Figure 158 below), and only by taking some of the twist off the rope-springs could the 100 degrees be achieved. Remember that the final 5.5 centimetres of our *cheiroballistra*'s pullback to 70 degrees requires a force of 27.4 kg. The shorter inswinger arms are at a massive disadvantage, being short levers attempting to wind an extra 30 degrees onto the rope-springs. In the end the state of the inswinger at 90 to 100 degrees appeared to be so dangerous that I banned further tests at that pullback.

From our own practical experience of the *cheiroballistra* with an arm arc of 70°, and the evidence of the violence of its frame judder on the release of the missile recorded by the slow-motion camera (Wild Dreams TV *Ancient Discoveries*: 'Ancient Warfare'), anyone attempting to construct an inward swinging arms *cheiroballistra* or Hatra catapult with 100 degrees of arm arc, must ensure that the greatest possible safety precautions are taken against the potential explosion of parts. At the start of our testing the inswinger launched the two arms and bowstring as well as the missile, creating three missiles and a potential strangling noose as used by the Thuggees! We question the long-term reliability of framework put under such a strain.

The Morgan-Feeley Hatra quarter scale models as inward and outward swinging

Above. Michael Lewis's models of old and new style catapults side by side. Top, arms at rest. Bottom, ready to shoot. Experiments with the two models resulted in an overall 50% higher performance by the 'inswinger'.

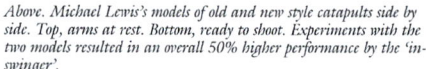

Figure 155 - Mike Lewis' models

As Figure 155 shows, Mike Lewis is comparing an outswinging Xanten-Wardt (left) with a wide-frame inswinging Hatra catapult. Frame sizes are different. This is not comparing like with like. (Anonymous 2004, 48)

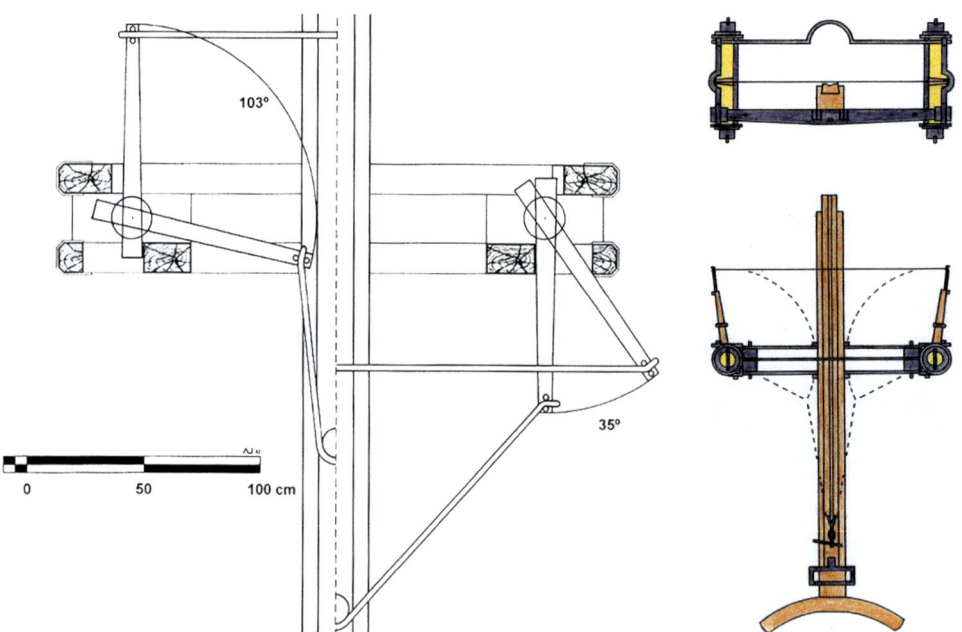

Figure 156 - Hatra inswinging designs: (left) Aitor Iriarte's two versions of the Hatra catapult. (right) Mike Lewis' inswinging cheiroballistra.

Figure 157 - Tom Feeley's quarter-scale inward swinging Hatra catapult

Tom spent a year modifying the superb, strongly-constructed model in Figure 157 in an attempt to make it work. I can assure those who would accuse the three of us of bias against the inswinger, that we tried our best to succeed. The overall length of the bowstring is 46 cm, half that of the Morgan-Wilkins outswinger at 91.5cm. The outswinger arms are 31.5cm, the inswinger's are 23cm. The outswinger bowstring travels 57.5cm along the slider, the inswinger only travels 40cm along the slider. See the comments on the characteristics of the inswinging design in Section 2 above.

So, although the evidence quoted above, in particular the way the Hatra catapult has landed, appears to rule it out as an inswinger, we decided to use the surviving evidence to produce both outswinging and inswinging arms versions at a quarter scale. Our inswinger is based on Aitor Iriarte's diagram. Very small-scale models such as those by Mike Lewis (Figure 155) allow only limited conclusions to be drawn. In any case Mike's small models of an outswinging Xanten bolt-shooter set up against an inswinging Hatra design, claiming 50% more power for the latter, are not comparing like with like (Anon. 2004, 48)

I have always given full consideration to alternative views, and have been intrigued by inswingers since a discussion in 1978 with Dietwulf Baatz, who, as an archer, was

Figure 158 - Tom Feeley's inswinging version in three positions (photographs: Mark Hatch)

not happy about the looped bowstring configuration; hence some of my remarks (*JRMES* 2000, 100) on the possible characteristics of the quite different geometry of inswinging arms and this bowstring.

In the top photograph in Figure 158 the arms are pulled back through nearly 90°, the furthest we were able to achieve safely: see text above. Note the tightly squeezed cup on the bowbelt. It was very difficult to feed the missile into this. In the bottom left photograph the arms are fully forward after missile release. The slider in the bottom right photograph has been moved forward to allow the trigger to be locked onto the bowbelt. Mark Hatch points out that on the full-size machine it would require a ladder to access the bowbelt and trigger because they would be well over 2.48m above ground level.

All three of our quarter-scale models used the same size of washers, were armed with the same volume of rope, and stretched onto the frames with the same tension. When tested with a 94g round projectile the following ranges were achieved. Vitruvian *ballista*: 35.7m; Hatra Inswinger: 47.6m; Hatra Outswinger: 52.53m. One interesting result from this is the fact that spacing the position of the Outswinger's spring-frames 2.25 times further apart than those of the Vitruviuan *ballista*, has resulted in an increase in power and range comparable to that achieved by the wide frame of the *cheiroballistra/manuballista* over the Vitruvian scorpion.

Figure 159 - A battery of twelve bolt-shooting and one stone-throwing catapults, June 2013

To conclude, the widespread modern pursuit of the inward swinging arms theory has tended to become an exercise *in vacuo*, losing sight of the artillerymen's battlefield experience and priorities. The time taken to wind back the suggested extra 30° of arm arc of an inswinger would slow down the rate of missile launch considerably: on the evidence of our tests it would have taken about 30% longer to load. The problem would have been compounded by the fact that the very much shorter arms provided reduced leverage on the spring bundle. No artilleryman would thank you for all of that. On the battlefield it was a question of rapid reloading as much as marksmanship to ensure survival. The fact that artillerymen were constantly seeking to increase the rate of launching (I am trying hard to avoid the modern phrase 'rate of fire'!) is proved by the introduction of the double ratchet push pull system shown on the Cupid Gem (Figure 85) and then the even faster crank handle system (Figure 88). While the surviving fragments of Hadrian's speeches to the army of Africa do not mention artillery, it is fascinating that rapid reloading is the subject of the emperor's comment in one of the sections recovered by Mark Speidel, (Speidel 206, 12, Field 22): Hadrian emphasises the need to practise rapid shooting with the bow: '*...when the enemy is close, your prefect pushes you to strive to shoot with greater speed and urgency, so that because of the density* [my underlining] *of the arrows the enemy does not dare to lift his head above his shield*'. Finally, Hadrian's governor-general Arrian's famous quote emphasises volume: '*It is expected that the indescribable volume of missiles will stop the charging Scythians from coming too close to our infantry line.*'

Our interpretation of the existing evidence is that the field artillery arch strut metal frame design, as described in the *Cheiroballistra* text and depicted on Trajan's Column, had outward swinging arms. Catapults with inward swinging arms may have been tested in the later Roman Empire and/or Byzantine period, but clear, incontrovertible evidence for this is needed.

It can also be argued that there was no requirement to increase the range of bolt-shooters, because they could already reach 500m or more.

Chapter 15

Survival

The one-arm *onager* is the only type of stone-thrower known to Vegetius, the late fourth century writer on the Roman army. This implies that it had replaced the Vitruvian and Hatran two-arm stone-throwers, whose complexity was one of the causes of their decline. They required much larger volumes of the sinew rope and the power of their two springs had to be critically adjusted until it was identical. The *onager* in its turn is unlikely to have survived until mediaeval times because it was eventually replaced by the traction trebuchet (Figure 148) which may have arrived from China by the end of the sixth or beginning of the seventh century AD. The demise of the single rope-spring powered *onager* was no doubt hastened by the readily available man power and the simple design and low cost of the traction trebuchet.

However, torsion bolt-shooters of the arch strut type continued in use. There is a fifth century AD inscription from Cherson (Crimea) mentioning *ballistarii*, *ballista* mechanics or artillerymen. *Polemikai hamaxai meta cheirotoxobolistron* ('war wagons with hand-arrow-shooters' - note the last word's close resemblance to *cheiroballistra* - were used by the people of Cherson in the late third and fourth centuries. *Carroballistae* are mentioned in tactical manuals of the seventh (Maurice's *Strategicon*) and tenth centuries (Leo's *Tactica*). The powerful psychological effect of the bolt-shooter's accuracy continued to be attested. Procopius records an incident in the siege of Rome by the Goths in AD 536 when a tower-mounted catapult shot a bolt which passed through the breastplate and body of a distinguished Gothic warrior and sank to over half its length in a tree. The sight of his corpse pinned to the tree caused utter panic amongst the Goths and they stayed out of missile range. The Byzantine historian John Scylitzes describes how a bolt-shooter defending a Byzantine camp *c*. AD 1050 saved the day when the leader of a tribe besieging the camp was run through and impaled on his horse *katapeltiko belei*, 'with a bolt from a catapult'.

So, torsion bolt-shooters, invented in the mid fourth century BC, were still used by Byzantine armies *c*. AD 1050, a remarkable lifespan of fourteen centuries. The arch strut type of bolt-shooter, in use by AD 102, had a lifespan of nine, or possibly more centuries.

Chapter 16

The Roman achievement

It has been claimed that the Romans made few improvements to the Greek artillery designs that they adopted. That this is untrue has been demonstrated above, first in the case of Vitruvius' *ballista*, whose reshaped spring-frames and special washers increased the power of the earlier version described by Philon. Second, a case has been made for the Hatra stone-thrower being a Roman army replacement of the Vitruvian machine with a new wide spring-frame design permitting increased arm travel and power, and a clear field of view for accurate aiming.

Third, Marsden was right to list the *cheiroballistra /manuballista* as a new category of bolt-shooter, and not as another version of the Hellenistic weapon, because this arch strut catapult and enlargements of the type are not simply a replacement of wood by metal, but a bold new design of frame, arms and stand. The 'safety' of the well-tried Hellenistic plated wood version was abandoned, and the challenge undertaken to create entirely in metal a frame whose characteristics produced increased power and much improved handling on the battlefield. The enormous stresses to which the metal frame would be subjected would have been unpredictable and potentially dangerous, implying a thorough programme of strain testing during the development stage.

A second charge levelled against the Romans is that they never attained the mechanical standards of the Greeks (Schambach 1883, discussed by Marsden 1969, 174 following). The author of the *Cheiroballistra* text is Greek, and the expertise of Greek artillery engineers continued to be consulted. However, the new frame, arms and stand must surely have evolved from the Roman artillerymen's demands for a reappraisal of user requirements and improved weapon characteristics. The success of this design of bolt-shooter must be credited to them and to the superb quality of Roman metalwork produced by the legionary *fabricae* and by the specialist workshops of Autun (central France), Trier (west Germany) and other centres set up to manufacture *ballistae*. It also exposes the folly of Frontinus' pessimistic view (*Strategems* III, preface), apparently expressed about the time when the arch strut catapults were at the design or even trials stage, that the development of engines of war had long ago reached its limit.

In 399 BC engineers working for Dionysius of Syracuse had invented a weapon, probably the *gastraphetes*, which outstripped the range and impact of hand-bows and slings. The invention of the rope-spring torsion catapult followed in the same century, and remained a power source unique to the western world. In the east the Chinese continued to rely on and develop highly efficient crossbows of many designs and sizes, with some armies fielding 10,000 crossbowmen or more (e.g. in 341 BC: Man

2007, 62). Their enormous siege catapults, using windlass-operated double or triple multiple bows, were capable of very long ranges, reputedly up to about 1,000 metres, but with the sacrifice of mobility and speed of operation.

At the end of a long line of development of torsion artillery, the compact and portable *cheiroballistra/manuballista* and its larger brothers represent one of the two most impressive and efficient designs of bolt-shooters in the history of western warfare, the other being the mediaeval military crossbow with its immensely powerful steel spring (Payne-Gallwey 1903, Chapter 5 etc.).

Chapter 17

Future search and research

The chances of finding Vitruvius' or Philon's lost diagrams, or new Greek or Latin texts related to artillery, are extremely remote. It is quite possible that information may survive in the countless unpublished Arabic copies of Greek or Latin texts. What are badly needed now are many more archaeological finds, preferably of complete assemblies of catapult parts like that of the Xanten-Wardt frame. This is the type of hard evidence required to achieve more accurate reconstructions, and to find answers to debated questions like the existence of inward swinging arms.

During the 250 years since the publication of William Newton's edition of Vitruvius and the early research by Köchly and Rüstow and de Reffye and Dufour, countless boltheads and *ballista* balls and dozens of washers have been found. However there have been a mere handful of major frame parts from the wood frame Vitruvian bolt-shooters: the iron frame cladding from Ampurias and Caminreal, the Cremona battle shield, and - at last - the virtually complete front frame of the Xanten-Wardt scorpion. Tantalisingly, the rear of the case and slider of the Xanten-Wardt is missing. Of the later arch strut bolt-shooter, we only have one large arch strut, several field-frames and washers, and a possible iron arm. All are individual parts; there is not even a pairing of arch with field-frame, let alone a complete front frame assembly. No complete case or slider has survived from either type of bolt-shooter. No trigger mechanisms or windlasses have been identified, or any stand. The word 'identified' is a reminder to keep on searching existing museum stores, remembering that the Lyon frame and washers and the Cremona shield remained misidentified for many decades. Of the Vitruvian/Heron *ballista*, not a single part has been identified. It is probable that it was superseded by the more powerful Hatra design of stone-thrower with its wide-spaced spring-frames echoing those of the *cheiroballistra*.

Even when new parts or assemblies of parts are found, it should be remembered that there would have been many variations in the details of parts for the same type of catapult. They were not produced by machines, and depended on the skills and materials available in each workshop. For example, there was probably more than one way of locking the parts of the *cheiroballistra/manuballista* together. The solution I have suggested may not be authentic. In particular the chunky field-frame from Sala (Volubilis, Morocco) may have required a different type of junction to the Arch and Ladder.

This is a gloomy picture. It looks as though we are going to have to wait for purely chance finds of assembled parts, discovered in exceptional circumstances like that at

Xanten-Wardt, rather than in archaeological excavations. Of the latter, sites in Britain like Vindolanda and Carvoran Fort could provide the best chances.

Marine archaeology offers some hope, as with the Mahdia shipwreck finds (Figure 34). Priority should be given to underwater search for catapults on warship wrecks, for example from the battle of Actium, where large numbers of *ballista* balls have already been located on the seabed by The Actium Project. There are also the smaller scale battles, such as that between Caesar and Pompey in the harbour at Brindisi where Pompey used large merchant ships converted with three storey towers packed with artillery to attack Caesar's harbour mouth blockade of rafts with two storey towers (Caesar *Civil War* I, 26).

As for experiment and research, the priority must surely be to investigate and produce sinew rope. Digby Stevenson deserves great credit for making a sufficient length of it to fit the *cheiroballistra*, though he reconstructed the machine as a stomach bow (Harpham and Stevenson 1979). Much experiment will be needed to manufacture sinew rope with the right breaking strain to withstand the required pretensioning and torsion. An alternative would be to design and manufacture synthetic rope with similar characteristics to sinew.

Tests with slow-motion cameras will improve understanding of the performance of bolts and stones as they travel up the sliders. Different cross-sections for the missile groove on the slider are worth trialling. I have often wondered whether the Romans lined the missile grooves with polished sheet metal to reduce friction and wear and increase range.

We would like to record that all testing of our reconstructions and all public demonstrations have been with the rope-springs on light tension, for obvious safety reasons. This should be noted when assessing the penetration results shown in Figures 51 and 52.

Appendix A

The Roman origin of the mediaeval revolving-nut crossbow release

Figure 160 - The fragmentary Roman hunting relief from St Marcel

Figure 161 - The Solignac hunting relief

The important stone relief in Figure 160, now in the Musée Croatier at Le Puy-en-Velay, France, shows a hunting party setting out into a forest, with a dog and mules to carry the prey. This is probably a hunt for deer. The lithe hound contrasts sharply with the powerfully-built Molossian hound in the second relief in Figure 161. The non-torsion crossbow in Figure 160 is carried under the huntsman's left arm along with a large quiver for the bolts. This allows the size of the crossbow to be estimated from the length of the huntsman's arm. I believe that this type of bow is the *arcuballista* (bow catapult) described by Vegetius IV.22 as used by *arcuballistarii* in battles alongside *manuballistarii*, slingers and archers. The thickness of this and the Solignac bow below suggests that they were very powerful composite bows. (photographs: by kind permission of the Musée Croatier).

This Solignac relief, also in the Musée Croatier at Le Puy-en-Velay, France, shows a similar hunting crossbow from above. The revolving-nut release for the bowstring is clearly portrayed, along with the groove for the bolt. It proves that the Roman crossbow is the ancestor of all mediaeval crossbows which use the same revolving-nut release. Also on the relief is a quiver and what seems to be a hunting knife in a scabbard. The accompanying heavily-built, muscular hound, probably a Molossian, would have been powerful enough to tackle a wild boar. It appears to be wearing a protective throat collar. (photographs: by kind permission of the Musée Croatier).

Appendix B

Review of 2021 TV film on Burnswark Hill, 'Massacre on Hadrian's Wall'

A programme for the American Smithsonian Channel and Channel Five in Britain.

This was first transmitted in America on the Smithsonian Channel in August 2021 under the title "*Rome's Secret Weapon*". It has been broadcast in Britain repeatedly from November 2021 for Channel Five under the '*Massacre on Hadrian's Wall*' title, and in 2022 in Australia on the SBS Channel. It is currently available on sites like ClickView. Some brief extracts are on YouTube.

Background information

In August 2020 my colleague Len Morgan and I were informed that the Smithsonian Institute, Washington and Channel Five TV channels wanted to make a film about reconstructing the *cheiroballistra/manuballista* catapult for a TV series on ballistics. In the end we agreed that to save the very high cost and time involved we would dismantle our master reconstruction and rebuild it. A team from Blink Films, London shot footage over three days in Annan, covering the story of the many years of international attempts to understand the famous five-page manuscript, the building of our reconstruction, and live shooting with the catapult at various targets including a large block of ballistic gel. The Blink Films team was impressive and they had done their homework on the subject. We understood that their footage would be sent up to a quite different team for editing and inserting into the final programme for the existing TV Series *Ancient Treasures*.

After a year's delay I was informed that the programme had aired in America on August 21, 2021 under the title *Rome's Secret Weapon*. A brief extract from this appeared on YouTube with John Reid launching and fronting the film as an account of the Roman army's massacre of native Britons on Burnswark Hill. Following the shock, I asked for a complete copy of the version about to appear on TV in this country. It arrived as the same version as had been transmitted in America, but with the different title *Massacre on Hadrian's Wall*. This is not a film on ballistics, and Len and I feel that we were misled. Had we known the actual theme of this film we would not have taken part.

The following critical appraisal uses a transcript of the exact words used by John Reid in his running commentary. It also refers to additional commentary voiced by a narrator, much of which mirrors John's words or contains remarks that John had already made in personal communications to me. I have assembled the quotations in groups and numbered them for easy reference.

1. [The screen picture shows Burnswark hillfort from above **with a mass of roundhouses and with its rampart complete, not collapsed**]

JR: '*Hillforts...they are usually heavily defended. They had ditches and ramparts all the way round. This one had triple ditches on the south side. They were quite clearly making sure that the defences were strongest there...*' **Narrator:** '*For the local tribes this would have seemed an impregnable fort*'. '*This was the most heavily defended hillfort in the region and home to thousands of native Britons...*'

Commentary

This screen picture showing upstanding ramparts contradicts the archaeological evidence for the collapse or slighting of the hillfort's rampart which Jobey's trenches proved. My photo of the catapult ball found in David Devereux's excavation of Trench 2 for Reid and Nicholson (above Figure 129) shows that a period of time had elapsed between the collapse or slighting of the rampart and the launching of the *ballista* ball found in this trench.

One of the most disturbing statements made by Reid in the whole of this TV broadcast is that this hillfort '*had triple ditches on the south side. They were quite clearly making sure that the defences were strongest there...*' They provenly do not exist. Reid has invented them. The text above my photograph in Figure 139 reads, '*David Devereux is showing Lawrence Keppie the stone facing of the Iron Age rampart tumbling down the hillside. This has been removed in the left half of the trench to investigate possible earlier features. None were found. The hillfort has no ditches, as both Barbour and Jobey's excavations had already proved (Barbour 1998, Plate IV; Jobey 1978, Figure 4).*'

2. JR: '*You are looking at a population up here of 2,000 maybe 2,500 people, and at a time of crisis that number would swell to maybe three times that.*'

Commentary

Jobey's limited area of excavation on the summit is the only archaeological search to date for houses still occupied during the second century there. He found one where the Roman pottery does give dates up to the mid 2nd century. Therefore Reid's Figure for the '*population up here of 2,000 maybe 2,500 people..*' is a totally unfounded guess. Its confident delivery preceded by the phrase '*You are looking at a population ...*' gives it an air of authenticity. Reid is introduced in the film as *Dr Reid, Roman historian*' which implies that his academic qualification is in this subject, which it is not.

'*...at a time of crisis that number would swell to maybe three times that.*' This is a further wild guess and is part of his imagined scenario of a major uprising. See my section on the 'Occupation of the hillfort after the collapse/demolition of the ramparts' on p.164.

Review of 2021 TV film on Burnswark Hill, 'Massacre on Hadrian's Wall'

3. JR: *'The army was coming in as a major battering ram. The assault was likely to have been a short, sharp, lightning attack which would have lasted probably only a few hours. ... This was a blizzard, a death-dealing blizzard.'* **Narrator:** *'... Lollius ordered his men to leave the safety of Hadrian's Wall, and 6,000 soldiers surged northwards into the wild lands beyond...Within hours of arriving Lollius had employed classic battle tactics and had the settlement surrounded.' 'Up to 4,000 soldiers took up position, while another 2,000 were positioned directly behind the fort.'* **JR:** *'The heavily armed troops disgorged out of the three big gateways down here and started advancing up the hill. It would have been a terrifying sight...'*

Commentary

This glosses over the fact that the missile bombardment took place after a detailed rebuild of the South Camp's NW rampart and ditch. Subsequently slingers launched lead bullets from this heightened rampart top, losing many which rolled back down the slope from there. The rebuild, which included cutting the bottom two feet of the ditch in solid rock, would have taken far more than '*a few hours*'. Reid envisages a '*lightning attack*' of the *veni vidi vici* type and yet he talks of the '*major military engineering*' that preceded the bombardment. He makes no mention of the heavy paving in the South Camp and at several gateways in both camps, evidence for repeated use of the site.

Lollius Urbicus. Of course, he must be considered a possibility as the legate of Britain when the Roman army's operations at Burnswark took place. Page 166 lists the governors that were in office during the timespan now suggested by the provisional Carbon 14 dating.

I have never been convinced by Reid's argument that because Lollius was one of Hadrian's generals in the war against the Jews when the Romans massacred countless Jews, his character must have been one liable to transfer such treatment onto the Britons. The weakness of this tendentious argument is revealed when it is realised that it could be extended and imply that all the generals under Hadrian's command in that war must have had this vicious streak. The fact is that we have no evidence of Lollius' personality.

Roman reaction to rebellions was in proportion to the relative strength and threat posed by the enemy. Only against formidable rebellions, such as that organised on a massive scale by the militarily well-equipped Jews, did Roman generals turn on the full power of the Roman army. The heavy Roman death toll achieved by the Jews was matched by the Romans' ruthless reaction.

4. JR: *'There would have been absolute mayhem here, men, women and children, animals, all packed into a very tight space. It would have been absolute butchery, and nobody would have stood a chance. The army was coming in as a major battering ram and bringing with them equipment that the natives would not have seen before. This was a site of a major military* *engineering exercise. The size of it, the magnitude, to me means only one thing: this had imperial backing. The spread of missiles, the number of missiles and the mix of missiles tell us that the Romans were throwing everything at the hilltop at the one time. We know it was at the one time because they are all in the same archaeological strata. We know that there are stone balls, lead bullets and iron boltheads all coming in simultaneously.'*

Commentary

The scene of butchery conjured up in Reid's emotional language has no evidential basis.

'The army was... bringing in equipment that the natives would not have seen before'. This is untrue and untenable. By the mid 2nd century memories may have become dimmed of the warfare during Venutius' revolt and the campaigns of Cerialis and the Ninth Legion. On Hadrian's succession the *Historia Augusta* records that *'the Britons could no longer be held under Roman control'*. The suppression of this British insurrection, commemorated by coins of 119, was achieved by Pompeius Falco, governor from 118 to 122. Marcus Aurelius' tutor Fronto recalled the great number of Romans killed by the Jews and Britons during Hadrian's reign. It is possible that the Vindolanda tombstone (RIB 3364) of Titus Ann[ius] centurion of the First Tungrians, *'in bell[o inter]fectus'*, *'killed in war'* or *'killed in the war'*, dates from this period.

It is regularly argued that there were challenges to the construction of Hadrian's Wall, when the legions and their artillery would have become an all-too-familiar sight

Presumably Reid makes this point to justify his belief that Britons were prepared to make a stand on the hill because they were quite unaware of the weapons the Roman army would field against them, in particular catapults.

5. Narrator. *'Something caused the Roman army to leave their impregnable line of defence – Hadrian's Wall. For 15 years the Romans held firm behind Hadrian's Wall. In AD 139 they pushed north and headed into the wilds that lay beyond the hard barrier of the Roman Empire. So why did they leave the security of Hadrian's Wall?'*

Commentary

See also the quotations from JR and the Narrator in 3 above.

The idea that during Hadrian's reign the Romans stayed behind the Wall and did not venture into the 'wilds' beyond would be laughable if it were not so horrendously far from the truth. It ignores the countless activities beyond the new frontier, such as the establishment of the Hadrianic forts at Birrens, Netherby and Bewcastle and the work on related road systems. As to the 'wilds' beyond the Wall, well-trodden by several generations of Romans, the census of the Anavionenses under Trajan soon after AD

Review of 2021 TV film on Burnswark Hill, 'Massacre on Hadrian's Wall'

100 had already added to the membership of the Roman Empire the very group of people who were neighbours to, if not the inhabitants of Burnswark Hill.

There are many other errors in this disastrous film. John Reid told me when I first met him that he had had a lifetime fascination with Burnswark as a scene of conflict between Romans and Britons. Over the years in a series of public lectures he has become increasingly assertive and self-convinced of this interpretation of the evidence. In this film, tricked out with repeated emotion-rousing scenes of male and female 'Britons' killed by Roman missiles, he has inflicted on TV audiences on both sides of the Atlantic, and now in Australia, a version of events which is almost entirely fictional. The few factual sequences include those in Dumfries Museum, where he and Andrew Nicholson display and analyse the sling bullets and other missiles, and the catapult sequences provided by Len Morgan and myself.

Finally, I record the fact that when I stood up at the end of Andrew's lecture on Burnswark in Tullie House Museum in November 2021 and asked about the film, he was taken aback and said that he had never heard of it. He had already told the audience that one of his jobs was to rein in John Reid. At the end of question time he said that in their forthcoming final report he would ensure the accuracy of the description of the excavations and their follow up actions and research, but could not be entirely responsible for the last section offering an interpretation of the results.

Massacre on Hadrian's Wall, the title itself a geographical error, continues to air regularly on TV internationally, with internet advertising claiming it as '*A shocking new discovery reveals just how ruthless the Roman army could be in pursuit of absolute power. This program reveals the brutal tactics behind Rome's blitzkrieg – the Empire's lightning war.*'

Reid's film is not simply a question of his different interpretations of the evidence. He has adopted an approach which ignores vital information and details of the site proven by his own Reid and Nicholson excavations and those of Barbour and Jobey. This approach involves several examples of wilful misrepresentations, such as the invention of ditches that did not exist and the failure to record that the defences were no longer upstanding. He makes the erroneous claim that the local Britons were unfamiliar with Roman weaponry. Because only one roundhouse has so far been proved to have been occupied at the time, his statement that there were maybe 2,000 to 2,500 Britons living on the top and that this number would maybe triple in a crisis has no factual basis and is entirely guesswork.

Readers and viewers will draw their own conclusions. The result of this film is that he has let down his audiences and his colleague Andrew Nicholson. By one error alone, inventing the existence of a major archaeological feature, he has done irreparable, fatal damage to his attempts to establish himself as a credible Roman historian.

Sources and references to artillery

Philon of Byzantium. Possibly active in the last half of the third century BC. He cites his long association with the Alexandrian artillery technicians and his contacts with many master craftsmen of Rhodes. His works include *Paraskeuastika* ('Siege preparations') and *Poliorketika* ('Siege-craft'). His *Belopoiika* ('Artillery Construction'), translated by Marsden, contains measurements for parts of the Hellenistic stone-thrower which can often be used to supply or correct missing or corrupt measurements in Vitruvius' *ballista* text. Unfortunately his diagrams, like those of Vitruvius, have not survived.

Vitruvius. His *de architectura* ('On architecture and engineering'), published *c.* 25 BC, is the only surviving work on the subject and has had a profound influence on world architecture. He was a catapult specialist and, along with three others, was put in charge of the construction and repair of artillery by Emperor Augustus, having also served Julius Caesar, probably in the same capacity. Therefore, his description of the bolt-shooting *scorpio* and stone-throwing *ballista* must represent the standard versions used by the early Imperial army. His diagrams have not survived. The version of his *ballista* text published by Schramm often strays some way from the manuscript evidence, and the present author has revised the texts for both machines.

Heron of Alexandria. One of the great Alexandrian engineers. He witnessed the eclipse of the moon in AD 62 and is assumed to have worked in the second half of the first century AD, possibly surviving into the second century. His *Belopoiika* ('Artillery Construction'), translated by Marsden, contains a review of the development of artillery from stomach-bow to torsion weapons, and much catapult lore on building and setting up the catapults, with particular emphasis on the stone-thrower. His diagrams do survive, but he does not give measurements for parts (unless he put this information in a section which has not survived). The *Cheiroballistra* text is not in his style of Greek and is unlikely to have been written by him.

Josephus. His eyewitness account of the Roman war against the Jews is translated with parallel Greek text by H. St J. Thackeray in Volumes 2 and 3 of the Loeb Classical Library edition of Josephus, Heinemann and Harvard, 1927. The 1959 translation by G.A.Williamson, *Josephus. The Jewish War*, is published in the Penguin Classics series. He translates the word for scorpions as 'quick-firers'.

Arrian. General and historian. Appointed by Emperor Hadrian as governor of Cappadocia (east Turkey), he blocked the invasion by the hordes of Alani in AD 134. His detailed order of march and briefing for the battle with the Alani survive (Chapter 11 page 145).

Vegetius. His *epitoma rei militaris* is the only manual on the Roman army to have survived intact. It was written sometime after AD 383, but is a scissors-and-paste assembly of material from various earlier periods in the history of the army. The fact that the only stone-thrower mentioned by him is the *onager* (Chapter 13) suggests that some of his remarks on the use and battle positions of artillery weapons are valid for the later Roman army. There is an English translation by N.P. Milner, *Vegetius, Epitome of Military Science* Liverpool University Press, 1993.

De Munitionibus Castrorum. This is the accepted title for this part-surviving important work on the construction of Roman camps. The unknown author was believed to be Hyginus Gromaticus, and is often referred to as Pseudo-Hyginus. Its date is still in dispute, but may be from the late first to early second century AD. That it does in part reflect actual practice is occasionally confirmed, for example by the large marching camp at Reycross on the top of the Stainmore Pass, where the document's advice to station traverses 60 feet in front of camp gates is largely adhered to (Richmond and McIntyre 1934, 54). Various editions of the Latin text are available online. The 1887 edition by von Domaszewski is printed and translated by Catherine Gilliver in the *Journal of Military Equipment Studies* 4, 33-48.

Accounts of Roman artillery in action

The scope of this book only allows the inclusion of a few of the accounts of artillery in action from the works of ancient authors. For these the reader should consult the relevant sections of Dr Eric Marsden's first volume**, Historical Development**, particularly Section III *The spread of artillery in the west*, Section V *Artillery in sieges*, VII *Catapults in field campaigns and naval warfare*, and VIII *Roman imperial artillery*. Marsden does not often translate his Latin and Greek quotations. Most sources are translated in the Loeb Classical Library series, many in the Penguin Classics series.

Dr Duncan Campbell's **Besieged** (Campbell 2006) is recommended for detailed accounts of Roman siege warfare. His accounts are illuminated by his personal knowledge of the sites in the Near East, most of which are unapproachable because of present day warfare. His dissertation for a Doctorate **(**Campbell 2002**)** contains a detailed analysis of the evidence for Roman sieges of all periods.

Select bibliography

Most of the important articles are scattered in specialist journals, but they can be found in certain major British libraries and some university collections. The *Saalburg Jahrbuch* is in many of these libraries and is also available online, as are a limited number of the following articles. Some of the volumes of the *Journal of Roman Military Equipment Studies*, usually abbreviated as JRMES, are available within the UK from The Armatura Press Amazon merchant store via the JRMES Shop page or, for the EU or the USA, from Karwansaray Publishing.

Aarts A. C. 2012: 'Scherven, schepen en schoeiingen. Archeologisch onderzoek in een fossiele rivierbedding bij het castellum van De Meern. *Basisrapportage Archeologie 43*. Utrecht.

Anonymous 2004: *Trajan's Artillery. Archaeology of a Roman Technological Revolution.* Current World Archaeology Number 3, 41-48. Presents Mike Lewis' inswinging interpretation of the *cheiroballistra*.

Arbiv K. 2023: 'Evidence of the Roman Attack on the Third Wall of Jerusalem at the End of the Second Temple Period'. *Atiquot* 111, 2023.

Baatz D. 1978: 'Recent finds of ancient artillery', *Britannia* 9, 1-17. The publication of the Hatra, Gornea and Orşova finds.

Baatz D. 1980: 'Ein katapult der Legio IV Macedonica aus Cremona', *Römische Mitteilungen* 87, 283 - 299. The full publication of the Cremona shield, washers etc..

Baatz D. and Feugère M. 1981: 'Éléments d'une catapulte Romaine trouvée à Lyon', *Gallia* 39, 201 - 209. The Lyon spring-frame and washers.

Baatz D. 1985: 'Katapultteile aus dem Schiffswrack von Mahdia (Tunesien)'. *Archäologischer Anzeiger*, 679-691. Some of these underwater finds are in Figure 34.

Baatz D. 1994: *Bauten und Katapulte des römischen Heeres*. Stuttgart. Reprints all of Baatz' artillery articles, and adds new chapters.

Baur P.V.C. and Rostovtzeff M.I. (editors) 1931: '*The excavations at Dura-Europos. Preliminary Report of the Second Season of Work, 1928 to 1929*'. New Haven Conn. Yale University Press.

Bishop M.C. and Coulston J.C.N. 2006: *Roman Military Equipment*. Batsford. Many illustrations and references to artillery. Second Edition 2006, Oxbow Books.

Birley A. 2002: *Garrison Life at Vindolanda*. Tempus, 114-116 for the account mentioning sinew.

Birley R. 1996: *Vindolanda Research Reports. New Series Volume IV. The Small Finds. Fascicule1. The Weapons.* The Vindolanda Trust.

Bosman A.V.A.J. 1995: 'Velserbroek B6-Velsen1-Velsen 2.' *Journal of Roman Military Equipment Studies* 6, 89-98.

Bosworth A.B. 1977: 'Arrian and the Alani' *Harvard Studies in Classical Philology* 81, 217-255.

Breeze D. 2011: 'Burnswark: Roman siege or army training ground?' *Archaeological Journal* 168, 166-180.

Breeze D. 2019: '*Oppida* in Britain in the face of the Roman conquest.' in Romankiewicz et al. (ed) *Enclosing space, opening new ground*, 121-128. Oxbow.

Bronson B. 1986: 'The making and selling of wootz, a crucible steel of India'. *Archeomaterials* 1, 13-51.

Brouquier-Reddé V. 1999: 'L'équipement militaire d'Alésia d'après les nouvelles recherches', *JRMES* 8, 277-88.

Burns M. 2004: 'Pompeii under siege: a missile assemblage from the Social War', *JRMES* 14/15, 1 - 9

Campbell D.B. 1984: 'Ballistaria in first to mid-third century Britain', *Britannia* 15, 75-84. Discussion on the identification of artillery emplacements in forts at Hod Hill, High Rochester etc..

Campbell D.B. 1986: 'Auxiliary Artillery Revisited', *Bonner Jahrbücher* 186, 117-132. Was artillery ever used by non-legionary troops? Discussions on Hatra and High Rochester etc..

Campbell D.B. 2002: *Aspects of Roman Siegecraft*. Thesis for the degree of Ph.D., University of Glasgow.

Campbell D.B. 2003: *Greek and Roman Artillery 399 BC – AD 363*. Osprey.

Campbell D.B. 2003: 'The Roman Siege of Burnswark' *Britannia* 34 (2003), 19-33.

Campbell D.B. 2006: *Besieged*. Osprey. Detailed accounts of Roman siege warfare.

Campbell D.B. 2011:'Ancient catapults. Some hypotheses re-examined', *Hesperia (Journal of the American School of Classical Studies at Athens)* Volume 80, 677-700. 677-685 contain a valuable analysis of the evidence for the invention of the catapult and the arrival of torsion artillery. He rightly refutes Rihll's suggestion that sling bullets were originally invented for catapult ammunition.

Campbell D.B. 2012: 'Hillfort under attack. Roman siege warfare or training exercise at Burnswark', *Ancient Warfare* V-6, 2-7.

Christison D, Barbour J and Anderson J 1899: 'Account of the excavations of the camps and earthworks at Birrenswark Hill .....', *Proceedings of the Society of Antiquaries of Scotland* IX, 195-249.

Collingwood R.G. 1925-26: 'Burnswark Reconsidered', *Trans. Dumfriesshire and Galloway Natural Hist. and Antiquarian Society* 13, 46-58.

Connolly P. 1981: *Greece and Rome at War*. Macdonald. Sections shared with Dr Roger Tomlin on the later Roman Army and Siege Warfare.

Dufour G. H. 1840: *Memoire sur l'Artillerie des Anciens et sur celle du Moyen Age*. Paris and Geneva. Available as Gyan Books reprint.

Eastern Dumfriesshire. 1997. RCAHM Scotland. 179-182 discusses Burnswark.

Foley V., Palmer G. and Soedel W. 1985: 'The Crossbow', *Scientific American* 252.1, 80-86.

Gudea N. and Baatz D. 1974: 'Teile spätrömischer ballisten aus Gornea und Orşova', *Saalburg Jahrbuch* 31, 50-72. The full details of the Gornea and Orşova finds.

Handelsbank B. 1989: *Römische Tecknik in Bayern*. Not stated.

Hardy R. 1992: *Longbow* (Third Edition). Patrick Stephens.

Harpham R. and Stevenson D. 1979: 'The Manufacture of Sinew Rope', *Journal of the Society of Archer-Antiquaries* 40, 13 – 17. The first successful recreation of catapult sinew-rope in modern times.

Hart V.G. and Lewis M.J.T. 1986: 'Mechanics of the *onager*', *Journal of Engineering and Mathematics* 20, 345-365.

Hassall M. 1999: 'Perspectives on Greek and Roman catapults', *Archaeology International* 1998-1999, 23-26.

Holder P. 1987: 'Roman Artillery 1', *Military Illustrated* 2, 31-37.

Holley A. E. 1994: 'The Ballista Balls from Masada', *Masada IV: The Yigael Yadin Excavations 1963–1965.*

Final Reports. 347-365. Israel Exploration Society. Jerusalem

Holley A. E. 2014: 'Stone projectiles and the use of artillery in the siege of Gamla', *Gamla III: The Shmarya Gutmann excavations 1976-1989, finds and studies: part 1*, 35-56. Israel Antiquities Authority, Jerusalem.

Howard-Davis C. 2009: '*The Carlisle Millennium Project. Excavations in Carlisle 1998-2001*' Oxford Archaeology North.

Iriarte A. 2000: 'Pseudo-Heron's *cheiroballistra* a(nother) reconstruction: I. Theoretics', *JRMES* 11, 47-75. Reconstruction as a hand-held, hand-spanned weapon, based on the ms reading of 11/3 dactyls (25mm) for the rope-spring diameter.

Iriarte A. 2003: 'The Inswinging Theory', *Gladius* XXIII, 111-140.

James S.T. 1983: 'Archaeological evidence for Roman incendiary projectiles', *Saalburg Jahrbuch* 39, 142-3.

James S. T. and Taylor J.H. 1997: 'Parts of artillery projectiles from Qasr Ibrim, Egypt', *Saalburg Jahrbuch* 47, 93-98.

Jobey G. 1977-8: 'Burnswark Hill', *Transactions of the Dumfriesshire and Galloway Natural History and Antiquarian Society* 53, 57-104. Excavations on the hillfort and Roman camps, and a summary of the missile finds.

Jones R. H. 2011: *Roman Camps in Scotland*. Society of Antiquaries of Scotland. 154 – 156 on the Burnswark camps.

Jones R. H. 2012: *Roman Camps in Britain*. Amberley Publishing. 20-24 on Burnswark.

Kayumov I. and Minchev A. 2010: 'The Kambestrion and other military equipment from Thracia' *Proceedings of the VII Roman Military Equipment Conference, Zagreb* 327-342.

Keppie L. 2009: 'Burnswark Hill: native space and Roman invaders', in W.S. Hanson (ed) *The Army and Frontiers of Rome*. Journal of Roman Archaeology Suppl. Ser., 74, 41-52.

Keppie L. 2023: *Slingers and Sling Bullets in the Roman Civil Wars of the Late Republic, 90-31BC*. Archaeopress.

Landels J.G. 1978: *Engineering in the Ancient World*. Chatto & Windus. 99-132 discusses the engineering principles of ancient artillery.

Lepper F. and Frere S.S. 1988: *Trajan's Column*. Alan Sutton. 105-7 for the artillery scenes.

Man J. 2007: *The Terracotta Army*. Bantam Press.

Manning W.H. 1985: *Catalogue of the Romano-British iron tools, fittings and weapons in the British Museum*. British Museum Publications Ltd. Hod Hill boltheads in 169-177 and Plates 82-5.

Marsden E.W. 1969: *Greek and Roman artillery. Historical Development*. Oxford. Reprinted by Sandpiper, 1999. Indispensable, masterly survey of the historical background and defensive systems.

Marsden E.W. 1971: *Greek and Roman artillery. Technical Treatises*. Oxford. Reprinted by Sandpiper, 1999. Excellent editions, with English translations and discussion, of the Greek texts of Heron and Philon. Marsden's early death prevented his intended revision of the sections on Vitruvius and the *Cheiroballistra*.

Maxwell G. 1998: *A Gathering of Eagles. Scenes from Roman Scotland*. Historic Scotland. Includes a reassessment of the camps at Burnswark.

Miks C. 2001: 'Die ΧΕΙΡΟΒΑΛΛΙΣΤΡΑ des Heron', *Saalburg Jahrbuch* 51, 153-233. Reconstruction as hand-spanned, following Baatz.

Newton W. 1780: *Commentaires sur Vitruve: une description des machines militaires*. London. Available as Ecco reprint.

Newton W. 1791: *The Architecture of M. Vitruvius Pollio*. London (for James Newton).

Payne-Gallwey R. 1903: *The Crossbow (with … an Appendix on the Catapult, Balista and Turkish Bow)* London.

Prou V. with Vincent M. 1862: *La Chirobaliste d'Héron d'Alexandrie*. Paris. Available as Hachette reprint.

Prou V. 1877: 'La Chirobaliste d'Héron d'Alexandrie'. *Notices et Extraits…de la Bibliothèque Nationale*, xxvi, 2, 1-319.

RCAHMS. 1997: *Eastern Dumfriesshire an archaeological landscape*. HMSO.

Reid J.H. 2016: 'Bullets, ballistas and Burnswark', *Current Archaeology 316*, 20-26.
Reid J.H. and Nicholson A. 2019: 'Burnswark Hill: the opening shot of the Antonine reconquest of Scotland?' *Journal of Roman Archaeology 32*, 459 – 477.
Richmond I.A. and McIntyre J. 1934: 'The Roman Camps at Reycross and Crackenthorpe'. *Transactions of the Cumberland and Westmoreland Archaeological Society.* Series Two, xxiv, 50-61.
Richmond I.A. 1935: 'Trajan's Army on Trajan's Column', *Papers of the British School at Rome* XIII, 3-40. Reprinted with an introduction by Mark Hassall, 1982. *British School at Rome.*
Richmond I.A. 1962: 'The Roman army and Roman religion.' *Bulletin of the John Rylands Library*, Manchester, 45.
Richmond I.A. 1968: *Hod Hill Vol. II: Excavations carried out between 1951 and 1958*, London. 32-3 and Figure 14 for the artillery barrage.
Rihll T. 2007: *The Catapult. A History*, Yardley, Pennsylvania.
Russell M. 2019: 'Mythmakers of Maiden Castle. Breaking the siege mentality of an Iron Age hillfort.'
Oxford Journal of Archaeology 38, 325-342.
Schalles H-J et al 2010: *Die Frühkaiserzeitliche Manuballista Aus Xanten-Wardt.* Xanten Berichte Band 18. Verlag Philipp von Zabern. A detailed account of the painstaking efforts to conserve this highly important archaeological find. The description of the *scorpio minor* catapult as a *manuballista* is erroneous. Baatz's interpretation as a hand-held, hand-spanned weapon is unconvincing.
Schneider R. 1906: 'Herons *Cheiroballistra*', *Mitteilungen des Deutschen Archäologischen Instituts, Römische Abteilung* XXI, 142-168.
Sharples N.M. 1991: *English Heritage Book of Maiden Castle.* Batsford.
Schramm E. 1917: 'Erläuterung der Geschützbeshreibung bei Vitruvius X 10-12'. *Sitzungberichte der königlich preussischen Akademie der Wissenschaften, phil.-hist. Klasse.* LI, 1917, 718-734.
Schramm E. 1918: *Die antiken Geschütze der Saalburg.* Reprinted 1980 with an introduction by D. Baatz. Saalburg Museum. An indispensable classic from one of the great exponents of experimental archaeology.
Sim D. and Ridge I. 2002: *Iron for the Eagles. The iron industry of Roman Britain.* Tempus. Second Edition 2012. Sim's authoritative experiments on the manufacture of weapons.
Sim D. and Kaminski J. 2012: *Roman Imperial Armour.* Oxbow Books. Definitive research into all technical aspects of the production of this armour.
Spiedel M.P. 2006: *Emperor Hadrian's speeches to the African Army – a new text.* Mainz 2006. The best edition to date.
Stefan A. S. and Chew H. 2015: *La Colonne Trajane.* Picard Paris 2015. Reproduces at large scale the original photographs of the Napoleon III casts of the Column.
Stiebel G.D. and Magness J. 2007: 'The military equipment from Masada'. *Yigael Yadin Excavations 1963-1965 Final reports.* Israel Exploration Society. Jerusalem 2007.
Tittel K. 1913: Pauly-Wissowa, *Real-Encyclopädie* 1913, 1039-41. Rightly challenges Schneider's strange verdict on the *Cheiroballistra* manuscript.
Todd M. 1978: *The Walls of Rome.* Paul Elek. 87 BC catapult emplacement in Figure 4, and other references to artillery on the later walls
Ucelli G. 1940: *Le Navi di Nemi.* Istituto Poligrafico e Zecca dello Stato, Rome.

Ureche P. 2013: 'The Bow and Arrow during the Roman era'. *Ziridava Studia Archaeologica* 27, 183-196.
Vicente J.D. *et al* 1997: 'La catapulta tardo-republicana…de 'La Caridad' (Caminreal, Teruel)'. *JRMES* **8**, 167-199.
Wescher C. 1867: *Poliorcétique des Grecs.* Imprimerie Impériale. Paris
Wheeler R.E.M. 1943: *Maiden Castle, Dorset.* Oxford UP for the Society of Antiquaries of London.
Wilkins A. 1995: 'Reconstructing the *cheiroballistra*', *Journal of Roman Military Equipment Studies* 6, 5-59. Text, translation and reconstruction as a winched bolt-shooter.
Wilkins A. 2000: '*Scorpio* and *cheiroballistra*', *Journal of Roman Military Equipment Studies* 11, 77-101. *Scorpio* reconstructed from revised Vitruvius text and finds of parts. Update on the *cheiroballistra*.
Wilkins A. 2002: 'The Graeco-Roman stone-throwing catapult', *Timber Framing* Number 65, 18-19. With articles on building and lifting the one talent BBC *ballista* by Gordon Macdonald and Professor Grigg Mullen Jr..
Wilkins A. 2003: *Roman Artillery*, Princes Risborough: Shire.
Wilkins A., Barnard H. and Rose P.J. 2006: 'Roman Artillery Balls from Qasr Ibrim, Egypt' *Sudan and Nubia* Bulletin **10**, 66-80, and version with additional illustrations as a pdf on the Roman Military Research Society's website 2011: http://romanarmy.net
Wilkins A. and Morgan L. 2011: 'A suggested reconstruction of Vitruvius' Stone-thrower: *de Architectura* X, 11, 4-9', *Journal of Roman Military Equipment Studies* 18, 27-50. pdf on the Roman Military Research Society's website 2011: http://romanarmy.net
Wilkins A. 2011: *The Xanten-Wardt Roman catapult, and catapult parts from Carlisle.* pdf on the Roman Military Research Society's website 2011: http://romanarmy.net

Museums and sites

Museums

Many museums in the area of the Roman Empire display artillery boltheads or stone balls. The following is a selective list of those museums with the most important artillery finds or reconstructions, most of them on display. Please check telephone numbers and opening hours before visiting: I cannot guarantee that the following information is up-to-date or that a particular item is on public display.

BRITAIN

British Museum, Great Russell Street, London WC1B 3DG. Telephone 020 7323 8299. Vindolanda Tablet II, 343 the 'sinew' document (page 60-1). Boltheads (Figure 19) and military equipment from Hod Hill in the Roman Britain gallery. Our Xanten-Wardt bolt-shooter (page xi) is displayed in the Museum's 2024 Exhibition *Legion: Life in the Roman Army.*
Roman Army Museum, Vindolanda Trust, Carvoran Roman Fort, CA8 7JB Telephone 01697 747485. There is an impressive artillery display of Tom Feeley's Xanten-Wardt and *manuballista* bolt-shooters.

SOURCES AND REFERENCES TO ARTILLERY

Dewa Roman Experience, Pierpoint Lane, Off Bridge Street, Chester, Cheshire CH1 1NL. Telephone 01244 343407. Small working model *onager*.

Dorset County Museum, High West Street, Dorchester, Dorset DT1 1XA. Telephone 01305 262735. The two famous Maiden Castle victims (Figure 8). The square hole in one skull was almost certainly caused by a Roman *pilum* thrust, not a catapult bolt.

Dumfries Museum, The Observatory, Dumfries, Dumfries and Galloway DG2 7SW. Telephone 01387 253374. Burnswark stone shot and sling bullets.

Great North Museum: Hancock, Barras Bridge, Newcastle on Tyne, NE2 4PT. Telephone 0191 208 6765. The High Rochester inscriptions and the large catapult balls are on display (see below under the site).

Hunterian Museum, The University of Glasgow, Glasgow G12 8QQ. Telephone 0141 330 4221. Stone shot and composite bow nocks from Bar Hill Roman fort (page 125).

National Museums of Scotland, Chambers Street, Edinburgh EH1 1JF. Telephone 0131 225 7534. Stone shot, arrow heads and sling bullets from Burnswark.

Ribchester Roman Museum, Riverside, Ribchester, Preston PR3 3XS. Telephone 01254 878261. Stone shot and boltheads on display.

Roman Baths Museum, Pump Room, Stall Street, Bath, Somerset BA1 1LZ. Telephone 01225 477773. Small bronze washer from a scorpion (Figure 34), spring diameter 41 mm, possibly thrown as an offering into the sacred spring, perhaps by an engineer hoping for Sulis Minerva's help.

Royal Armouries at Fort Nelson, Fort Nelson, Fareham, Hants PO17 6AN. Telephone 01329 233 734. Sir Ralph Payne-Gallwey's large *onager* (Section 14) and Eric Marsden's *cheiroballistra* (Figure 58).

Trimontium Museum at Newstead. Market Square, Melrose TD6 9PN. Telephone 01835 342788. Len Morgan's *carroballista* (Figure 91), three-span bolt-shooter and quarter-scale *ballista*.

Tullie House Museum and Art Gallery, Castle Street, Carlisle, Cumbria CA3 8TP. Telephone 01228 534781. The two Carlisle catapult frame parts (Figure 41), my 1991 initial version of the *cheiroballistra* (Figure 60) and small working model *onager*; boltheads, and stone shot and sling bullets from Burnswark.

Vindolanda Fort and Museum, Chesterholm Museum, Bardon Mill, Northumberland NE47 7JN. Telephone 01434 344277. Display case of catapult boltheads, many of the size for the Xanten-Wardt scorpion.

BULGARIA

Nova Zagora Historical Museum, Tsentar, ploshtad Svoboda 3, 8900 Nova Zagora, Bulgaria. Telephone 0889 123 456. Holds the Kambestrion and other 1962 finds published by Kayumov and Minchev 2010.

FRANCE

Musée de la Civilisation Gallo-Romaine, 17, Rue de Cleberg, 69005 Lyon. Telephone 04 72 38 49 30. The Lyon field-frame (Figure 48).

Musée des Antiquités Nationales, Chateau – Place Charles de Gaulle, 78100 St-Germain-en-Laye. Telephone +(33)1 39 10 13 00. Victor Prou's reconstruction of the Cheiroballistra (Figure 36). De Reffye's scorpion, ballista and onager. These are in the reserve store at the time of writing.

MuséoParc Alésia, 1, route de Trois Ormeaux, 21150, Alise-Sainte-Reine. Telephone +33 (0)3.80.86.96.23. Display of Roman missiles from Caesar's siege (Figure 18). A reconstruction of a section of the Roman siege lines (similar to Figure 17). Unfortunately, the museum's scorpion reconstruction has undersized washers and no plating.

GERMANY

Limesmuseum, St Johann Strasse 5, 73430 Aalen, Baden-Wurttemberg. Telephone 07361 528 287-0. A replica of the Caminreal frame (Figure 33).

Saalburgmuseum, Saalburg-Kastell, 61350 Bad Homburg vor der Höhe. Telephone 06175 9374-0. Those of Schramm's catapults that survived the Second World War are on display in the museum of the reconstructed fort.

LVR-RömerMuseum, Archäologischer Park Xanten, Trajanstrasse 10, 46509 Xanten. Telephone (for education and advice) +49 (0)2801 9889213. The Xanten-Wardt catapult frame (Chapter 5. 5) as envisaged by Baatz as hand-spanned.

ITALY

Museo Civico, Palazzo Affaitati, via Dati Ugolani 4, 26100 Cremona. Telephone 0372 407770. The Cremona battle shield (Figure 45).

Musei Vaticani, Città del Vaticano, Roma. Telephone +39 06 69883145. The Vedennius tombstone (Figure 35) is in the Chiaramonti Gallery (a statue of Diana stands on top of it).

Trajan's Column is in the Roman Forum. A pair of binoculars and/or a camera with a long telephoto lens are essential to pick out the artillery scenes on it. Watch out for pickpockets!

Museo della Civiltà Romana, Piazza G. Agnelli,10- 00144 Roma. Telephone 060608. This is at E.U.R., near the station EUR Fermi on Linea B of the Rome Metro). The complete set of casts of Trajan's Column are at eyelevel, but are broken up into short lengths.

THE NETHERLANDS

Castellum Hoge Woerd, Hoge Woerdplein 1, 3454 PB De Meern, Utrecht. Telephone 030-2860084. The one complete and two part-complete catapult bolts, and the *ballista* stones (Figures 3 and 115) will be displayed in one of the towers along with Len's reconstruction of the Ampurias bolt-shooter (Figure 32).

ROMANIA

National Museum of Romanian History, Bucharest, 12 Victoriei Avenue, District 3, Bucharest. Telephone +40 21 315 82 07. Permanent display on Trajan's Column and the Dacian Wars. As with the above E.U.R. display, the complete set of casts of Trajan's Column are at eyelevel, but are broken up into short lengths.

Sources and references to artillery

SPAIN

Museu Arqueològic Barcelona, Pg. de Santa Madrona 39-41, 08038 Barcelona. Telephone (+34) 93 423 21 49. The Ampurias catapult frame in a revised display (Figure 32).

Museo Provincial de Teruel, Plaza Fray Anselmo Polanco, Apartado de Correos 119, 44071 Teruel, Aragon. Telephone: 978 60 01 50. The Caminreal catapult frame, and washers from Azaila (Figures 33 and 34).

Sites in Britain

There are very few sites in Britain where artillery is known to have been in action, because there are no accounts of long sieges of impregnable strongholds like Jerusalem and Masada.

Sites from the Roman Invasion

Hod Hill, Dorset (Section 5). Ordnance Survey Sheet 194, G.R. 8510. One access is from the village of Stourpaine. The site of the 'chieftain's hut' is not marked, but its approximate position can be worked out from Richmond's article.

Maiden Castle, Dorset (Section 5). 2 km south west of Dorchester, off the A354. Ordnance Survey Sheet 194, G.R. 6788. It is worth visiting the Dorset County Museum in Dorchester first (above) to see the two skeletons.

The northern frontier

Burnswark Roman Camps, Dumfries and Galloway (Chapter 12). O.S. Sheet 85, G.R.1878. Access by the unclassified road which leaves the B725 by the railway bridge 1 km north east of Ecclefechan. After 3 km on this unclassified road, park at the T-junction (near the bottom left corner of Figure 126). Please respect the grazing sheep.

High Rochester Roman fort, Northumberland. Ordnance Survey Sheet 80, G.R. 8398. The two third century inscriptions from the site, RIB 1280 (Figure 144) and 1281, mention the repair of *ballistaria*, emplacements for catapults. They are on display in the Great North Hancock Museum (above) along with the large stone catapult balls weighing 24, 28 and 42 kg, the first two being suitable for a one-talent (80 *librae*) stone-thrower, the 42 kg shot implies the presence at High Rochester of a one-and-a-half talent machine.

Catapults on Roman defences

Chester Roman Fortress, South East Angle Tower (foundation courses visible from the wall by the Newgate). Bryan Sitch (pers. comm.) gives details of a 1.1 kg stone shot of red sandstone in the Manchester Museum, recorded as from this angle tower, further evidence of the siting of stone-throwers in a defensive role.

Later Roman coastal defences

Burgh Castle, Norfolk. O.S. Sheet 134, G.R.4704. Signposted from the roundabout on the A149, 3 km north of Great Yarmouth. The impressive remains of one of the Saxon Shore forts which protected the coast against Saxon raids in the late third century AD. The huge drum of the fallen bastion of the south wall has a round socket on the top which may be part of a base for a catapult, possibly a large bolt-shooter of the type and size of the Orşova arch (Figures 90 and 91). All these coastal forts would have been well equipped with both bolt- and stone-shooting catapults.

Websites and TV programmes

There are many websites offering information on Greek and Roman artillery, some of which display little connection to the ancient evidence. Only one or two re-enactment groups display authentic catapults. There is an illustrated section on artillery on the website of the Roman Military Research Society at www.romanarmy.net.

Some television programmes I have been involved in dealt with specific aspects of the subject. Few of them are now available. Some are on YouTube; unfortunately, they are usually in truncated extracts. Others may occasionally be repeated on programmes on the History Channel or the like. If they can be found, the following TV programmes contained relevant information:

What the Romans Did for Us, Episode 2 *Invasion*. BBC 2000.

Building the Impossible: The Giant Ballista. BBC 2002. Unfortunately, at that time I had not started my research into Vitruvius' *ballista*.

Ancient Discoveries Series: Ancient Warfare. 2005. Also *Ballistics.* 2008. Wild Dream Films, originally for the History Channel.

Beat the Ancestors: The Roman War Machine. 2013. Channel 5. At the time of publication there is an important video on YouTube: *Huge Roman Catapulta* by Tod's Stuff. This shows the four-cubit bolt-shooter from the TV programme, now equipped with authentic Roman arms, achieving ranges over 300m.

The highly unsatisfactory, historically inaccurate 2021 TV film *Massacre on Hadrian's Wall* is reviewed in Appendix B above. It does however contain the important shoot at ballistic gel (Figure 64) and other targets.

Sources and references to artillery

Where the reconstructed catapults described in this book can be seen

The majority of the reconstructions illustrated throughout the text have been realised in collaboration with his two highly skilled engineer colleagues Len Morgan and Tom Feeley. Several of their full-size reconstructions can be seen in museum displays on Hadrian's Wall at Carvoran Roman Army Museum, Tullie House Museum, Carlisle, Trimontium Museum at Newstead, Colchester Castle Museum, Park in the Past near Wrexham and the Castellum Hoge Woerd at De Meern near Utrecht. The *gastraphetes*, the two *librae ballista*, the *cheiroballistra*, two Xanten-Wardt bolt-shooters, the new Morgan-Wilkins version of the *onager*, and some scale models will have a permanent home in the forthcoming History Centre at Comlongon Castle, Dumfries, where they will be regularly displayed in action. Also regularly seen there will be the author's version of the Greek semi-automatic bolt-shooter, the *polybolos*.